Relativity, Supersymmetry, and Strings

Relativity, Supersymmetry, and Strings

Edited by

Arnold Rosenblum
Director, International Institute of Theoretical Physics
Utah State University
Logan, Utah

Plenum Press • New York and London

Library of Congress Cataloging in Publication Data

International Institute of Theoretical Physics School on Relativity, Supersymmetry, and Strings (1986)
Relativity, supersymmetry, and strings / edited by Arnold Rosenblum.
p. cm.

"Proceedings of the International Institute of Theoretical Physics School on Relativity, Supersymmetry, and Strings, held February 24–28, 1986, in Logan, Utah"—T.p. verso.
Includes bibliographies and index.
ISBN 0-306-42680-3
1. General relativity (Physics)—Congresses. 2. Grand unified theories (Nuclear physics)—Congresses. 3. Supersymmetry—Congresses. 4. String models—Congresses. I. Rosenblum, A., 1943– . II. Utah State University. International Institute of Theoretical Physics. III. Title.
QC173.6.I575 1986
530.1'1—dc19 87-21294
CIP

Proceedings of the International Institute of Theoretical Physics School on Relativity, Supersymmetry, and Strings, held February 24–28, 1986, in Logan, Utah

A Division of Plenum Publishing Corporation
233 Spring Street, New York, N.Y. 10013

Printed in the United States of America

PREFACE

The first international conference on relativity physics and parallel computing took place February 24-28, 1986, at the International Institute of Theoretical Physics at Utah State University in Logan, Utah. Before proceeding to summarize some of the main points of the conference I would like to make a few remarks about the goals of the International Institute of Theoretical Physics. The Institute undertakes frontier research in both pure and applied physics and is housed in the new physics building at Utah State University. The applied research is in solar energy and optical parallel computing. We are in the midst of building four very efficient solar collectors and, in the very near future, parts of an optical parallel computer in the Institute.

The pure research is concerned with relativity physics, particle physics, condensed matter physics, quantum systems, etc. On the macroscopic level, we are concerned with possible experimental tests of Einstein's General Relativity. On the microscopic level, we are interested in the unification of gravitation with the other interactions. In addition to research, the International Institute sponsors visiting scientists and encourages collaborations, etc.

To return to the conference, Professor Ehlers, Professor Rindler, and Professor Rosenblum represented the general relativity part of the conference. Professor Ehlers spoke about the problems connected with quantizing general relativity. The main point is that it is very difficult technically and conceptually to quantize a theory like general relativity which is both a field theory and Riemannian geometry at the same time. Professor Rindler gave a set of lectures concerning cosmological models. He emphasized the geometrical aspects of the models. Professor Rosenblum, after reviewing tests of general relativity using the binary pulsar, went on to discuss how one may use clocks for tests of both dragging of inertial frames and gravitational radiation. He ended his lecture with an introduction to Parisi-Wu stochastic quantum field theory with the intention of applying it to quantum gravity.

The next set of lectures concerned the unification of the interactions in particle physics. Professor Slansky began with an introduction to string theory, and Professor Raby gave lectures on supersymmetry phenomenology.

To represent the applied physics aspects of the International Institute of Theoretical Physics, Dr. Martin Walker gave a set of lectures on parallel computing. His company, Myrias Corporation, is in the midst of building one.

Shorter contributions were made by Professor Wu on "Quantum Curvature and Anomalies in Gauge Theory" and Zi Wang on "Strings and Cosmology." The participants found the conference very stimulating, as we hope the readers will.

Arnold Rosenblum, Director
International Institute of
Theoretical Physics

CONTENTS

WHERE STRINGS CAME FROM AND WHAT THEY ARE

Lecture 1

STRINGS AND THEIR COMPACTIFICATION

FROM THE PARTICLE VIEWPOINT

R. Slansky

T-8, Theoretical Division
Los Alamos National Laboratory
Los Alamos, NM 87545 USA

It is not possible to cover all the developments of string theories in four lectures. Moreover, not everyone in this audience is working on strings, which makes it difficult to select a level of presentation that might be interesting to the experts, and at the same time be pedagogically helpful to those who would like to learn some of the basics of string theory. Thus, it is necessary to restrict severely the topics, and I have done so with the nonexperts in mind. To indicate the level, I am told that I should not assume knowledge of the Nambu action [1], although many of you certainly know it in detail. So, in discussing general string formalism, I will omit technical details and many important topics and, instead, attempt no more than a qualitative description. When discussing compactification, I will describe the simplest toroidal schemes, including some of the Lie algebraic background. Nothing will be said about Calabi-Yau manifolds, which have been so interesting in the context of the heterotic string.

Further restriction of topics is a division of labor with Stuart Raby. He will discuss the progress in making a covariant field theory of strings, while I will emphasize the particle formulation. This latter formulation follows from the recognition that the modes of various relativistic strings generate the physical states appearing in the dual amplitudes. Thus, the string is a device for finding the Hilbert space of states. A prescription is then given for computing amplitudes. The vertex operators of this prescription depend on the string functions. The procedure has been checked in many ways to obey the fundamental axioms of a quantum field theory, but it is not derived from the underlying theory. Although it may look somewhat *ad hoc,* these prescriptions are the theory [2].

There are two classes of models that we will discuss. The first set are the old models that are associated with the dual resonance models. These include the open and closed bosonic strings, the Neveu-Schwarz model, the Ramond model and the open superstring model. These models all existed a decade ago. From the modern viewpoint, none of these models are directly useful, because the one-loop amplitudes are not finite, and because the field-theory limits have anomalies, or fail to include gravity. However, they are important building blocks in putting together even bigger theories that do have a great deal of consistency. There of five of these theories. They have the illuminating names, Type IIA, Type IIB, the $E_8 \times E_8$ and Spin(32)/$\mathbb{Z}_2$ heterotic strings, and the SO(32) Type I theory! These theories will receive some attention throughout the lectures.

String theories are generalizations of Yang-Mills theories, and closed strings also include gravity. Superstrings include fermions [3]. At present the amplitudes are availiable, and are believed to be the above generalizations because they are perturbatively equivalent to the theories they generalize. In the case of gravity, the generalization is somewhat richer, since the terms that lead to nonrenormalizable theories are replaced by terms where heavy particles are exchanged.

Like many, I have a strong preference for the efforts to make a field theory formulation of strings, since the S-matrix formulation does not provide a framework that is capable of answering many questions that field theories can. For example, one important issue concerns the symmetries of the ground state, which are simply put into the S-matrix theory by hand. The generalization of spontaneous symmetry breaking and the Higgs mechanism is not described in a natural fashion in the particle theory. More generally, the mismatch of the full underlying symmetry (of a Lagrangian) and the symmetry of the states is not easily seen in the particle theory. Thus, if the physical principle that underlies string theory is a generalized symmetry principle, then the S-matrix framework is surely an awkward way to find it.

This state of affairs should not appear unfamiliar, since there are historical precedents. For example, recall the connection between the amplitudes of the old Fermi vector theory of the weak interactions (or the V - A refinement) and the SU(2) X U(1) Yang-Mills electroweak theory. It was 30 years between Fermi's theory and the first suggestion of the SU(2) X U(1) theory, although the S-matrix elements given by the 4 - Fermi interaction gave excellent agreement with experiment. Of course, there were many efforts, including the one in Fermi's original paper, to model the weak interactions after electrodynamics. Such efforts led to many powerful results such as the conserved vector current hypothesis. Although the interaction Lagrangian of the V - A theory was written as such, it did not define a fully consistent quantum field theory, since the higher-order graphs derived from this Lagrangian did not make sense. It took another 10 years to clarify the standard model. No one would deny the importance of the underlying local SU(2) X U(1) symmetry of the standard model, especially since the discoveries of the W and Z bosons. However, that symmetry is most difficult to recognize from an analysis of a limited number of weak-interaction matix elements. The vacuum of the standard model has only a local U(1) symmetry of electrodynamics, so it takes a full set of amplitudes to recognize the full local SU(2) X U(1) symmetry.

Nevertheless, the particle approach can be defended, not so much in principle as in practice. So far, it has been so much easier to use than the field theory efforts, that it has been the only source of new theories. No new theories have been discovered in the field theory approach. Although less pleasing aesthetically, the particle approach is quite usable, and all the superstring theories with gravity were discovered using this approach. Moreover, when attempting to make a field theory of strings, researchers are constantly "peeking at the answers in the back of the book." If some effort to make a field theory yields amplitudes that do not agree with some dual model, it is has been taken as indicating an error. In formulating field theories, the particle theories are continually used for guidance. Of course, it is easy to be optimistic that someone will soon find a formalism so powerful that the field approach will finally become clearly superior. In the meantime, the particle formulation remains central to the concerns of string theory.

What is the problem in formulating string field theory? Why can't we just barge ahead and construct this theory? Part of the problem is a technical one. Lagrangian theories with manifestly independent degrees of freedom q^i and $\acute{q}^i$ are easier to quantize because of the successful prescription of turning the Poisson bracket into a quantum commutator; or, up to ordering ambiguities, the action can be used in a path integral without modification. However, in every fundamental theory of interest, there are *constraints* among the degrees of freedom that allow an elegant formulation of the theory. This is true of electrodynamics, Yang-Mills theories, and Einstein gravity and its generalizations. These theories cannot be quantized without paying proper attention to the constraints. It is true even of relativistic point particle mechanics, where the action is $\int ds$. (Most texts on classical mechanics make it a point to ignore such theories, even though it has been 35 years since Dirac made the first giant steps forward in understanding constrained Hamiltonian systems.) In the usual but old-fashioned approach, a direct effort to solve these constraints requires upsetting the manifest Lorentz invariance. For example, for a free relativistic point particle, the action is the length of the world line traced out by the motion of the particle. It is

$$S = -m\int ds = -m\int_{\tau_1}^{\tau_2} (-\acute{x}^\mu(\tau)\, \acute{x}_\mu(\tau))^{1/2}\, d\tau,$$

where $\acute{x}^\mu(\tau) = dx^\mu(\tau)/d\tau$. The canonical momentum is

$$p_\mu = \partial L/\partial \acute{x}^\mu = m\, \acute{x}_\mu/(-\acute{x}^2)^{1/2}$$

which obviously satisfies the constraint

$$p^\mu p_\mu + m^2 = 0\,.$$

The usual solution to the constraint is to solve the constraint directly by, for example, selecting $x^0(\tau) = \tau$. Clearly this violates the manifest Lorentz invariance, since the parameter τ is a Lorentz invariant notion. Actually, τ is not even a physical observable since the action can be "reparameterized." The symmetry operation $\tau \longrightarrow \tau'(\tau)$ leaves the form of the action unchanged.

A covariant field theory can be gotten simply by the substitution $p \longrightarrow -i\partial$, which gives the Klein Gordon equation. This trick is a little too fast for string theory, because the coordinates of the Nambu Lagrangian (which is a generalization of the point particle Lagrangian) for string theory satisfy an infinite number of constraints. In string field theory there must be some device for taking care of these constraints, since the argument of the string functional are the string coordinates.

The motion of a string is described in a way similar to the motion of a point particle $x^\mu(\tau)$. The motion of a string sweeps out a world sheet, which requires a function of two parameters $X^\mu(\tau, \sigma)$. In a gauge where τ is equivalent to time, $X^\mu(\tau, \sigma)$ (τ fixed) can be viewed as tracing out a line in space time. Thus, we select a parameterization where τ takes on all values, but $0 < \sigma \leq \pi$. An open string is one where $dX^\mu(\tau, \sigma)/d\sigma = 0$ at the endpoints $\sigma = 0$ and $\sigma = \pi$. Closed strings naturally satisfy $X^\mu(\tau, \pi) = X^\mu(\tau, 0)$, so that the world sheet is shaped like a cylinder in its motion between τ_1 and τ_2.

In the first efforts to make a string field theory, such as those by Kaku and Kikkawa or Cremmer and Gervais [4] in the early 70's, there was no suitable covariant formalism to deal with this difficulty, and almost all progress was made by solving the constraints directly. In this solution, the world sheet of the string $X^\mu(\tau, \sigma)$ was expanded in terms of independent modes of the transverse (to the light-cone) strings (plus a few more zero modes). Then the quantization could be carried out. (Part of today's discussion will be the identification of these modes.) Although the resulting formalism looks noncovariant, it is possible to prove the Lorentz covariance of all the particle amplitudes. It is also easiest to formulate the amplitudes in terms of these light-cone degrees of freedom. However, the beauty of the procedure is questionable, and it is not easy to approach many field theoretical problems while being stuck in light-cone gauge. (Just try to discover the Higgs mechanism in the light-cone gauge version of a Yang-Mills Lagrangian.)

Before we begin our survey of the particle formulation, let me tread for a moment longer on turf that Stuart Raby will cover, to help make it obvious that we are discussing the same theories. It has become possible in the past few years to covariantly quantize systems with constraints [5]. Recall that in transforming to independent degrees of freedom in the path integral, one must evaluate the Jacobian of the transformation. This is then lifted into the exponential where the action is using the ghosts and antighosts of the Faddeev-Popov procedure. It has proved possible to elevate this procedure for quantizing gauge theories to one that is physically appealing. The idea is that additional degrees of freedom can be introduced into the Lagrangian that will cancel the unphysical degrees of freedom of the original Lagrangian. This can be done in a manifestly Lorentz invariant fashion. Although this may sound a little silly for a classical system, it is not at all for the quantum theory. The quantum Lagrangian then has, not fewer degrees of freedom than the original Lagrangian, but more degrees of freedom, and the functional integral is done over more fields. In this formalism, referred to as "BRST" for Becchi, Rouet, Stora and Tyupkin, the ghosts and antighosts are related to the unphysical degrees of freedom of, for example, the gauge fields, by a supersymmetry. The terms of this Lagrangian include a gauge fixing term, so the local symmetry is violated explicitly in the Lagrangian, but is replaced by a global supersymmetry (generated by the BRST operator Q.) The new Lagrangian has no constraint equations, and so it is straightforward to write down the path integral. This way of getting a quantum theory is not too outrageous: The physical predictions require doing the path integral. All that is modified is the integrand and the variables that are integrated over. There can be many different integrals that give the same result. It is an amusing exercise to carry this procedure out for the relativistic point particle, and to derive the Klein-Gordon propagator directly.

Much of the effort over the last year has been sorting out the BRST formulation (including the Fradkin-Vilkovisky improvements [5]) of string theories. It is easy to be quite optimistic about the future of this business, and eventually about a complete string field theory based on some of these technical tricks. The noninteracting string particle theory has already been solved in this formalism [6] and the noninteracting field theory is in quite good shape. Raby will report on some very exciting progress in the use of strings as the arguments of field functionals. But the chore has not been completed for the interacting field theory, and so if we want to look at explicit models, it is necessary to turn to the dirty business of working in theories where the constraints are explicitly solved, and the manifest Lorentz invariance is lost.

Without an explicitly covariant formalism, it is more difficult to motivate the present formulations of string theories. The theories are like answers looking for a justification. One might suppose that there is some grand physical principle that should replace all the rules and procedures that surround the present particle apprach. However, the approach can be made somewhat more understandable by sketching the development of string theories. Hopefully, spending a few moments with the history [2] will make the newcomers to string theory a little more comfortable with the results that are the subjects of these lectures.

Twenty years ago there was a great ongoing effort to understanding hadronic scattering amplitudes from the consistency conditions of quantum field theory. The equations of the day, the so-called "bootstrap" equations, made several predictions that were not in accord with experiment. First, the width of the ρ meson was predicted to be about three times greater than seen experimentally. Secondly, the bootstrap equations suggested that the spins of hadronic resonances should be restricted to rather small values, smaller than had already been observed. The boostrap equations contained important assumptions, that two-body unitarity should be dominant, and that the double dispersion relations (that is, the Mandelstam representation) should have a very limited number of subtractions. The narrowness of resonances that were observed to lie on linearly rising Regge trajectories led to an alternate approach to hadron physics.

After a period of recognition of these problems, it was discovered that the Euler β function seemed to give a good first approximation to hadron amplitudes. This was discovered by Veneziano. Generalizations of the β function were rapidly discovered during a very active period of hadronic physics work. Of course, the formulas were required to have one-body unitarity (factorization of pole residues), Regge behavior, and they had the additional property of duality, which in this instance means that the s-channel poles of the rising Regge trajectories averaged out to look like a t-channel Regge exchange. This was a notable increase of efficiency over ordinary field theory, where s- and t-channel poles are added separately.

So, there were all these amplitudes. The next problem was to find out their meaning. In particular, the factorization showed there to be an enormous increase of multiplicity of particles of increasing mass. In fact it grew like exp(mass). There is a simple way to understand this remarkable increase in a more orderly fashion. The multiparticle dual amplitudes can be written in terms of integrals of matrix elements of certain combinations of harmonic oscillator operators. An infinite set of oscillators of the form, $a_n{}^{\mu}$, were introduced that satisfied the commutation relations,

$$[a_n{}^{\mu},\ a_m{}^{\nu}] \ = \ n\delta_{n,-m}\eta^{\mu\nu} \qquad \mu,\nu = 1 \text{ to } D.$$

(The normalization has become conventional, as it simplifies many formulas.) The Lorentz metric tensor $\eta^{\mu\nu}$ is a bit of a problem, since these operators define a Hilbert space that does not have a positive norm. Nevertheless, let us suppose that the mass-squared is related to the operator

$$N = \Sigma_{n=1}{}^{\infty}\ a_{-n}{}^{\mu}a_{n\mu}$$

Then we can count the number of states at each value of the N operator. All this is simple theory of the harmonic oscillator. In particular, the partition function is the generating function for this counting:

$$\sum_{n=0}^{\infty} M_D(n)x^n = \prod_{k=1}^{\infty} (1-x^k)^{-D}$$

The asymptotic behavior is $M_D(n)$ is $\exp(\mathrm{const}\sqrt{N})$.

Although the next step can be argued several ways, the main point is clear. There seemed to be some relation between the harmonic oscillators on a string and the states of the dual model. The only difficulty, which was an important one, was the appearance of negative norm states. In the old formalism, it was quite a chore to get rid of them, and other unphysical degrees of freedom.

The problem was to find a relativistic description of a line or string in space time that led to the same harmonic oscillator operators that could be used to make the physical states of the dual resonance model. After several clever attempts to do so, Nambu arrived at the most attractive string action [1]. He assumed that the relativistic string is described by an action that measures the area swept out by the world sheet of the string, just as the point particle is described by an action that measures the length of the world line. This may sound like a trivial generalization in words, but the Nambu action implied an infinite number of constraints on the world sheet, $X^\mu(\tau,\sigma)$. That there should be an infinity of constraints is not surprizing, if you just recall the point particle example.

A consistent solution can be obtained by solving the constraints implied in the classical theory with the Nambu action,

$$S = (1/4\pi\alpha')\int dA \quad ,$$

which is the area swept out by the motion of the string. The normalization constant α' has units $(1/\mathrm{GeV})^2$ (called the Regge slope), so S has the units of action when Planck's constant is unity. (We also set $c = 1$.)

The differential of area is just the magnitude of the cross product of dX^μ and dX^ν, which is just a 2 form,

$$dX^\mu \wedge dX^\nu = (\acute{X}^\mu X'^\nu - \acute{X}^\nu X'^\mu)\, d\tau \wedge d\sigma,$$

with $X'^\mu = dX^\mu/d\sigma$ and $\acute{X}^\mu = dX^\mu/d\tau$. Although this geometrical language is helpful for those who know forms, it should be emphasized that all we are achieving here is an explicit form of dA in terms of X^μ. The action is then

$$S = (1/2\pi\alpha') \int d\tau \int d\sigma\, [- \acute{X}^2 X'^2 + (\acute{X} X')^2]^{1/2}.$$

The constraints result from the reparameterization invariance,

$$\tau \to \tau'(\tau, \sigma)$$

$$\sigma \to \sigma'(\tau, \sigma).$$

It is not difficult to show that this Lagrangian implies

$$(\acute{X} \pm X')^2 = 0.$$

This is equivalent to an infinite number of constraints, as can be seen by taking the Fourier components of the constraints. The Fourier modes of the constraints, called L_n become the generators of the Virasoro algebra, which is the constraint algebra of the quantum theory.

So there is an infinity of first-class constraints. These can be solved in light-cone gauge, after which you may quantize the theory, as was done by Goddard, Goldstone, Rebbi and Thorn [7]. Of course, this is not the preferred order in which to do things, so the consistency of the procedure must be checked. Since light-cone formalism is not manifestly covariant, it is necessary to check that the quantized theory remains Lorentz invariant, which is true for the Nambu action only for $D=26$. Thus, the derivation of the physical states of the theory requires solving the constraints one way or another.

We can see from a naive model what can go wrong if these constraints are ignored. Suppose that each component of an open bosonic string field $X^\mu(\sigma,\tau)$ ($\mu = 0, \dots, D-1$) satisfies a wave equation with open string boundary conditions $\partial X^\mu/\partial\sigma = 0$ at the endpoints of the string, $\sigma = 0$ and π. (The world sheet swept out by the motion of the string is parameterized by σ and τ.) Suppose we substitute the constraint equations into the Nambu action. Then we can find that it is equivalent to the action for noninteracting waves,

$$S = \int d\sigma d\tau\, (\acute{X}^2 - X'^2)\ ,$$

which implies the wave equation,

$$\partial^2 X^\mu/\partial\tau^2 - \partial^2 X^\mu/\partial\sigma^2 = 0.$$

The solution to the wave equation with open-string boundary conditions is

$$X^\mu(\sigma,\tau) = x^\mu + 2\alpha' p^\mu \tau + \sqrt{2\alpha'}\, \Sigma_{n=1}{}^{\infty} (1/n)\, a_n{}^\mu \cos n\sigma\, e^{-in\tau},$$

where each mode appears to have D creation operators associated with it. The creation operators are $a_{-n}{}^\mu = a_n{}^{\mu+}$, $n > 0$. If the constraints are ignored, the canonically quantized string would be characterized by the commutation relations $[a_n{}^\mu, a_m{}^\upsilon] = n\, \eta^{\mu\upsilon}\, \delta_{m+n,0}$, since the solution to the free wave equation is just an infinite set of decoupled harmonic oscillators. The $a_{-n}{}^\mu$ operators acting of $|0\rangle$ in all combinations create an infinite dimensional Hilbert space. As mentioned before, this form is needed for a covariant theory, but it cannot be a consistent quantum theory, since the timelike oscillators would create states of negative norm. (We cannot ignore the constraints, as we already found from the Nambu action, so not all the oscillators are independent quantum operators.) In the quantized theory there are D - 2 oscillators associated with each mode. These are the light-cone bosonic string operators $a_{-n}{}^i$, where the index i runs over directions transverse to $x^\pm = (x_0 \pm x^{D-1})/\sqrt{2}$. The $a_n{}^+$ operators ($n \neq 0$) are set equal to zero in light-cone gauge, and the $a_n{}^-$ depend on all the ani operators. The relation is found simply by substituting these expansions into the constraint equations. Thus, the i index runs from 1 to D-2, and the metric in the transverse space is δ^{ij}. The set of quantization rules for the modes of the string are

$$[a_n{}^i, a_m{}^j] = n\ \delta^{ij}\ \delta_{n+m,0}\ ,$$

$$[x^i, p^j] = i\ \delta^{ij}$$

where the zero-mode momentum is related to the zero-mode oscillator by $a_0{}^i = \sqrt{2}\,\alpha'\,p^i$ and the zero-mode position operator is interpreted through the second commutation relation; it is the average position of the string. This restatement of the commutation relations is not unduly repetitive, since the oscillators in light-cone gauge are *not* given by the transverse components of the naive string operators, but by a more complicated relation.

The Lorentz generators M^{ij} for the transverse degrees of freedom are the obvious operators quadratic in $a_n{}^i$. The other generators are more complicated. In particular, the dependence of the $a_n{}^-$ operators on the ani's leads to a rather complicated computation, especially for the commutators involving M^{i-} ; it originally took a major effort and series of papers to straighten this out. One finds that the $[M^{i-}, M^{j-}] = 0$ only for $D = 26$ for the bosonic string. Generalizations including fermions are successful only in 10 dimensions, and finally, there does exist a $D = 2$ model. It is not very well understood whether these restricitions of dimensionality are due to the somewhat unsatisfying quantization procedure, or if they are truly intrinsic, although they reappear in covariant formalism. It is possible to avoid special values of D in free theories, but when local string interactions are included, the restrictions seem to reappear.

With this sketchy outline of the derivation of the quantum mechanics of the string, we turn to computing the string spectrum. This will require some knowledge of representation theory of semisimple Lie groups. We begin with the bosonic operators ani that describe the open bosonic string. Later we look at the spectrum of the closed string and will introduce the superstring fermionic operators $S_n{}^a$. Through all of this lecture, we regard the extra dimensions as flat, and consider compactification later.

The physical states lie in a Hilbert space constructed from products of a_{-n}^{i} (Fock space). The mass operator, up to a scale and an additive constant, is the number operator for the Fock space,

$$N = \Sigma_{n=1}^{\infty} a_{-n}^{i} a_{n}^{i}$$

The mass operator for the open bosonic string is just the constraint equation, which can be written,

$$\alpha' M^2 = N - 1.$$

We then define a product of operators $a_{-I}^{\{h\}}$ to be

$$a_{-I}^{\{h\}} = (a_{-1}^{i})^{N_1} (a_{-2}^{j})^{N_2} \ldots..$$

where $|I| = \Sigma_{n=1}^{\infty} n N_n$ is the number operator for the state created by a_{-I}. The helicity **{h}** are all the representations in $(v^{N_1})_S \times (v^{N_2})_S \times$ ...., where the subscript S means symmetrized part of the product and v is a vector in the SO(D-2) subgroup of the Lorentz group SO(1,D-1). The resolution of the state $a_{-I}^{\{h\}}|0\rangle$ into D dimensional spin consists of two parts. The massless helicity representations {h} are computed from the tensor product above. Then, for the massive fields the SO(D-2) irreps must be gathered into SO(D-1) irreps, which is the rest-frame symmetry for describing the spin of massive states in D dimensions.

In order to show some examples, let us suppose the a_{-n}^{i} are the bosonic operators in a superstring theory so that D = 10, Then the i index transforms as the $\mathbf{8_v}$ of SO(8).

The N=1 states are constructed as $a_{-1}^{i}|0\rangle = |i\rangle$, which transforms as a *massless* vector in D = 10. (When we include supersymmetry, the ground state is not a singlet $|0\rangle$, since by itself it does not form a representation of any supersymmetry.) It is time now to practice some of the group theory before working out the more complicated cases. The states for $N \geq 2$ are massive.

The spectrum of the open bosonic string spectrum for D=10 is:

N	Spin [SO(9) irreps]	Dynkin Description
2	**44**	(2000)
3	**36** + **156**	(0100) + (3000)
4	**1** + **44** + **231** + **450**	(0000) + (2000) + (1100) + (4000)

The light-cone gauge closed string is a solution to the transverse wave equation with boundary conditions $X^i(0, \tau) = X^i(\pi, \tau)$. Thus, there are two sets of nonzero-mode oscillator operators. The solution to the wave equation is then

$$X^i(\sigma, \tau) = x^i + 2\alpha' p^i \tau + i\sqrt{\alpha'/2}\, \boldsymbol{\Sigma}_{n \geq 0} (1/n)[a_n{}^i e^{-2in(\tau-\sigma)}$$

$$+ \tilde{a}_n{}^i e^{-2in(\tau+\sigma)}] ,$$

where an operator from one set of oscillators commutes with all operators in the other set. The condition that the origin of the string can be shifted around the string without physical implications imposes that $N = \tilde{N}$. Thus, if we begin with $|0\rangle$ as a ground state, then the first level (which is massless) is given by $a_{-1}{}^i \tilde{a}_{-1}{}^j |0\rangle$, which has the SO(8) irreps in $\mathbf{8}_v \times \mathbf{8}_v$.

More generally, the closed string states at level N are the tensor square of the open string states.

Note that for D = 10, the bosonic states made of bosonic operators yield the closed string spectra:

N	Symmetric Part	Antisymmetric Part	Algebra
1	**1** + $\mathbf{35}_v$	**28**	massless SO(8)
2	**1+44+450+495**	**36+910**	SO(9)

After these warm-up exercises, we are ready to add fermionic coordinates to the bosonic coordinates by supersymmetry. On the light-cone the supersymmetry is between the eight transverse bosonic degrees $X^i(\sigma,\tau)$ and the eight degrees of freedom for the open string $S^a(\sigma,\tau)$ of a Weyl-Majorana spinor in 10 dimensions. The free (noninteracting) string action in light-cone gauge is given by

$$S = (1/4\pi\alpha') \int d\sigma d\tau \,[-\partial_\alpha X^i \partial^\alpha X_i + i\alpha' S\gamma^- \rho^\alpha \partial_\alpha S] \quad ,$$

where ρ^α are the γ matrices in the 2-dimensional $\sigma-\tau$ space ($\rho^0 = I$ and $\rho^1 = \sigma_3$) and γ^- is a 10-dimensional γ matrix. Thus, the light-cone gauge action differs from a Dirac action by the insertion of the γ^-. It would not be satisfactory to omit the γ^-, even though it would apparently make the fermionic part of the string action Lorentz invariant: $\bar{S}\rho.\partial S$ vanishes identically because $S(\tau,\sigma)$ is a Weyl-Majorana spinor.

The solution to the light-cone Dirac equation $\rho.\partial S = 0$ with open-string boundary conditions, $S_1{}^a(\tau,0) = S_2{}^a(\tau,0)$ and $S_1{}^a(\tau,\pi) = S_2{}^a(\tau,\pi)$, is

$$S_1{}^a(\tau,\sigma) = \Sigma_{n=-\infty}{}^{\infty}\, S_n{}^a \exp\bigl(-in(\tau-\sigma)\bigr) \quad ,$$

$$S_2{}^a(\tau,\sigma) = \Sigma_{n=-\infty}{}^{\infty}\, \tilde{S}_n{}^a \exp\bigl(-in(\tau+\sigma)\bigr) \quad .$$

The index a on the fermionic operator is a weight of the $\mathbf{8_c}$ of SO(8). Canonical quantization gives

$$\{S_n{}^a \,,\, S_m{}^b\} \;=\; (\gamma^+(1\pm\gamma_{11})/2)^{ab}\, \delta_{m+n,0} \;.$$

and the bosonic and fermionic operators commute. For the open superstring, S and $\tilde{S}$ are the same, but for closed strings, they are independent operators, and an operator of the S type always anticommutes with an operator of the $\tilde{S}$ type.

The ground state of the open superstring is supersymmetric, and so it cannot be a singlet. It must have 10-dimensional N = 1 supersymmetry. Thus, the bosonic state $|i\rangle$ transform as the $\mathbf{8_v}$ of SO(8), and the fermionic states $|a\rangle$ must transform as the $\mathbf{8_s}$. Note that this is different from the $\mathbf{8_c}$ index on the fermionic operators. The zero-mode operator $S_0{}^a$ must transform the ground state bosons into the ground state fermions, and *vice versa*. Since $\mathbf{8_c} \times \mathbf{8_v}$ contains $\mathbf{8_s}$, it is necessary for the weights of $|a\rangle$ and $S_n{}^b$ to be in the two different congruency classes.

Finally, we need the mass operator for the open superstring, which is

$$\alpha' M^2 = N = \Sigma_{n=1}^{\infty} [a_{-n}{}^i a_n{}^i + (n/2)\, \overline{S}_{-n} \gamma^- S_n] \ .$$

Now we are ready to construct the excited states of the string, which is just a Fock-space construction as it was for the bosonic string.

The N = 1 bosons are constructed as $a_{-1}{}^i|j\rangle$ and $S_{-1}{}^a|b\rangle$, which transform as $\mathbf{8_v}^2$ and $\mathbf{8_s}\times\mathbf{8_c}$, respectively. The fermions are in $a_{-1}{}^i|a\rangle$ and $S_{-n}{}^a|i\rangle$. Thus, the bosons transform as a **44** + **84** of SO(9) and the first massive level of fermions are in the **128**.

The SO(9) content of the N = 2 and N = 3 levels of the open superstring is (the ground state has N = 0):

N=2 bosons **9+36+126+156+231+594**

N=2 fermions **16+128+432+576**

N=3 bosons **1+9+36+36+44+44+84+84+126+231+231+450+495+594+910+924+924+2457**

N=3 fermions **16+16+128+128+128+432+432+576+576+768+1920+2560**

The $N = 1$ level is an irrep of massive $D = 10$ supersymmetry.

The open string states fall into multiplets of $N = 1$, $D = 10$ supersymmetry. The open string theories are called Type I. In the case of the closed superstring theories, there are two sets of bosonic operators, and similarly, the boundary conditions on the Dirac equation for the $S^{\alpha a}$ allow for two independent sets of fermionic operators. The symmetry relations among these operators form an $N = 2$, $D = 10$ supersymmetry. (The closed string theories are called Type II.) This time there are two choices for the ground state. The ground state bosons are $|ij\rangle$ and $|ab\rangle$, and the fermions are $|ia\rangle$ and $|bj\rangle$. The question is whether the first and second fermion operators have the same set or a different set of weights a. In the Type IIA theories, one set transforms as $\mathbf{8_s}$ and the other set as $\mathbf{8_c}$. In this case the ground state bosons transform under SO(8) as $\mathbf{8_v}^2 + \mathbf{8_s 8_c}$ and the fermions as $\mathbf{8_v}(\mathbf{8_s}+\mathbf{8_c})$, which are also the states contained in the branching rules for the **44+84+128** of SO(9) into irreps of SO(8). Thus, the zero modes of this theory are the same as those of $D = 11$ simple supergravity (with a certain vacuum). In the Type IIB theory, the two sets of fermionic operators have the same SO(8) assignments.

The SO(8) assignments of the 256 ground state modes of the Type IIB superstring. We leave it as an exercise to compute the $N = 1$ and $N = 2$ spectra for both IIA and IIB. Note that the excited spectra of both closed-string theories are the same.

References

1. Y. Nambu, Lectures at the Copenhagen Symposium, 1970, unpublished.

2. For a grand introduction to dual resonance models, see the volume of Physics Reports edited by M. Jacob "Dual Theory" (North Holland, Amsterdam, 1975). Also see J. Scherk, Rev. Mod. Phys. **47** (1975),123.

3. For a review of superstring theory, see J. H. Schwarz, Physics Reports **89** (1982) 223 and M. B. Green, Physics Reports, to be published.

4. K. Kaku and K. Kikkawa, Phys. Rev. D**10** (1974) 1110; E. Cremmer and J.-L. Gervais, Nucl. Phys. B**90** (1975) 410.

5. For a review, see R. Marnelius, Acta Physica Polonica **B13** (1982) 669, or L. Baulieu, Physics Reports **129** (1985),1.

6. M. Kato and K. Ogawa, Nucl. Phys. **B212** (1983) 443; S. Hwang, Phys. Rev. D**28** (1983) 2614.

7. P. Goddard, J. Goldstone, C. Rebbi, and C. Thorn, Nucl. Phys. **B56** (1973) 109.

A QUICK INTRODUCTION TO SIMPLE LIE ALGEBRAS AND THEIR REPRESENTATIONS

Lecture 2

STRINGS AND THEIR COMPACTIFICATION FROM THE PARTICLE VIEWPOINT

R. Slansky

T-8, Theoretical Division
Los Alamos National Laboratory
Los Alamos, NM 87545 USA

Simple Lie algebras, their matrix representations and their subalgebra structure have long been studied by many mathematicians and physicists. Since physicists are interested in the Hilbert space of a quantum-mechanical theory, the traditional approach is to form the generators of a symmetry of a Lagrangian from the fields and then to analyze the symmetry directly in terms of the quantum operators. However, this approach is often cumbersome for dealing with the big groups that appear in string theory. It sometimes helps to proceed somewhat differently, similar to that popular in mathematics. There the focus is on the quantum number labels that appear in the state vectors [1].

The maximum number of simultaneously diagonalizable generators of a simple Lie algebra G is called its rank, ℓ; the total number of linearly independent generators is called its dimension dim(G). A simple group has no invariant subgroups, except for the whole group and the identity; analogously, a simple Lie algebra contains no proper ideals. A semisimple algebra can be written as a direct sum of simple algebras. Except for the study of subalgebras in the next lecture, we discuss semi-simple algebras only; U(1) is not simple.

In the standard Cartan-Weyl analysis, the generators are written in a basis where they may be divided into two sets. The Cartan subalgebra, which is the maximal Abelian subalgebra of G, contains the ℓ = rank(G) simultaneously diagonalizable generators H_i,

$$[H_i, H_j] = 0, \qquad i, j = 1, \ldots, \ell,$$

where the commutator [A, B] = AB − BA defines multiplication for a Lie algebra. The remaining generators are written so that they satisfy eigenvalue equations of the form,

$$[H_i, E_\alpha] = \alpha_i E_\alpha, \qquad i = 1, \ldots, \ell.$$

The numbers α_i are structure constants of the algebra in the Cartan-Weyl basis. For each operator E_α, there are rank(G) numbers α_i that designate a point in an ℓ-dimensional Euclidean space called root space. The term "root" refers to the fact that the root vector α is the solution to the eigenvalue problem given by the commutation relations. A fundamental problem of Lie algebra theory is to classify all possible root systems for algebras of each rank, consistent with the Jacobi identity, simplicity requirement, and the antisymmetry of the commutation relations. The dimension of the group is taken to be finite. The most elegant statement of the solution of these eigenvalue problems is given in terms of Dynkin diagram.

From the description of symmetries in quantum mechanics (or from Lie algebra theory), we recall that the generators H_i and E_α of G are characterized by their action of Hilbert space vectors, which describe the states of a physical system. A set of states that are necessarily interconnected by the E_α forms the basis of an irreducible representation (irrep) of G. There is an infinite number of finite-dimensional irreps. They are unitary for the compact real form of the algebra (where all generators are antihermitian when the group elements are exponentials of the generators without an i factor). The unitary representations of the noncompact real forms (where some generators are antihermitian and others hermitian) are all infinite dimensional and much more difficult to classify for the general case. The solution to the problem of finding all possible finite-dimensional irreps of any simple algebra is stated elegantly in terms of Dynkin diagrams.

The physical significance of the diagonalizability of the H_i is that the Hilbert space vectors $|\lambda\rangle$ in an irrep can be labeled by the rank(G) eigenvalues (or quantum numbers) of H_i : $H_i|\lambda\rangle = \lambda_i|\lambda\rangle$, $i = 1, \ldots, \ell$. Note that λ is not a complete set of labels, since it does not identify the irrep of G to which the set λ belongs, and also there are often several Hilbert space vectors in an irrep labeled by the same set $\{\lambda_i\}$, so further labels of the Hilbert space vectors are needed. The set $\{\lambda_i\}$ is called the weight of the representation vector. The solution to the problem of finding the complete list of weights of an irrep has a simple solution. The complete labeling problem, that of finding all the labels in addition to the weights, is more difficult, but a general solution will not be needed here.

The operators E_α are ladder operators, and they act just like the ladder operators of angular momentum theory. (In fact, the set E_α, $E_{-\alpha}$, and $[E_\alpha, E_{-\alpha}]$ for each α is an SU(2) subalgebra of G.) Suppose $|\lambda\rangle$ is an eigenstate of the H_i with eigenvalues (weights) λ_i, $i = 1, \ldots, \ell$. Then, $E_\alpha|\lambda\rangle$ is proportional to the state $|\lambda + \alpha\rangle$. It is necessary to learn when the proportionality constant is zero. Just as in angular momentum theory where the state $|j,j\rangle$ has the highest value of j_3 in the j irrep, there is always a highest weight vector of any finite dimensional irrep of G. We will see that this vector is very neatly designated in terms of the Dynkin diagram.

Let us finish writing down the commutation relations. If α is a root, then so is $-\alpha$. The commutator of E_α and $E_{-\alpha}$ is in the Cartan subalgebra,

$$[E_\alpha, E_{-\alpha}] = \alpha^i H_i,$$

where the components α^i are related to the α_i by a metric tensor.

The remaining commutation relations in the Cartan-Weyl basis have the form,

$$[E_\alpha , E_\beta] = N(\alpha, \beta)\, E_{\alpha+\beta} \qquad (\alpha+\beta \text{ not zero}),$$

where $N(\alpha, \beta) = 0$ if $\alpha + \beta$ is not a root. For example, 2α is never a root.

If we knew all possible ladder operators E_α (or more simply all possible root systems) for each rank, then we would know all possible simple Lie algebras, since the root vectors determine the structure of the Lie algebra. In the Cartan-Weyl basis, the nonzero roots of a simple algebra are nondegenerate; there is only one E_α for each α. The Cartan subalgebra may be viewed as being associated with an ℓ-fold degenerate zero root. The derivation of the root systems uses the commutation relations, the Jacobi identity, and clever manipulation to derive the crucial constraints on the roots. All possible root systems are summarized by the Dynkin diagrams.

It is helpful to identify a basis for the Euclidean root space, but this basis should be chosen with foresight and cleverness (or historical hindsight); the object is to avoid a mess in the general description of the generators and the representation theory. In the case of SU(3) the hypercharge Y and isospin projection I_3 are convenient only because there is no difficulty visualizing how these weights change under the U- and V-spin ladder operators. However, note that the U- and V-spin operators change both Y and I_3. Note that the hypercharge axis does not coincide with any SU(3) root. Thus, this simple Euclidean basis might be expected to be rather clumsy for higher rank groups, since the roots may point in somewhat complicated directions in a Euclidean basis. (This is only a difficulty in practice, but there is a way around it.)

For many purposes the best choice of basis of the root space is $\ell = \text{rank}(G)$ specially chosen linearly independent roots, because then the transformations induced by E_α have a simple description. A specially chosen set of roots, called simple roots, contain in a simple way all the information about the other roots and even about the weights of the representation vectors. The weights and roots are specified by the Dynkin diagram so conveniently that a detailed picture of weight space is unecessary for many purposes.

A set of simple roots can be found as follows. Write the roots in any Cartesian basis; half of the nonzero roots are positive, which is defined by the requirement that the first nonzero component of a positive root in that basis is positive. Then find the positive roots that cannot be written as a linear combination with non-negative coefficients of the other positive roots. There are only rank(G) such roots and they are linearly independent. That defines a set of simple roots. Of course, a different selection of coordinate system will lead to a different set of simple roots; however, the results on the relative lengths and angles of the roots are independent of the coordinate system.

The length and angle relations among the simple roots completely characterize a simple Lie algebra. Dynkin has shown how to summarize these relationships by a simple 2-dimensional graph, called a Dynkin diagram. Such a diagram must indicate the relative lengths of the simple roots and the angle between each pair of simple roots. Each simple root is denoted by a dot (or circle) on a diagram. In many algebras, all nonzero simple roots have the same length, so each root is designated by an open dot "O." No simple algebra has nonzero roots with three or more lengths. In those cases where nonzero simple roots have one of two different lengths, as is SO(5) and G_2, the longer root is denoted by O and the shorter root by a filled dot ●. It is convenient to normalize the length of the longer roots to $\sqrt{2}$.

The results of a somewhat lengthy analysis of the commutation relations and Jacobi identity reveals the following relations between lengths and angles. If two simple roots are orthogonal, there is no restriction on their relative lengths. If they subtend an angle of 120^0, the roots have the same length. If the angle is 135^0, the ratio of lengths is $\sqrt{2}$, and if the angle is 150^0, the ratio of lengths is $\sqrt{3}$. No other angles between simple roots are possible.

Since these restriction are so severe, it is easy to imagine that it is possible to classify the root systems in two dimensions. The rules are: if the angle between two dots (or simple roots) is 90°, do not draw a line between them; if the angle is 120°, draw one line; if the angle is 135°, draw 2 lines; and if the angle is 150°, draw three lines. If there is more than one line, the dot at one end must be open and the other one filled in. Returning to the SU(3) example, it is easily seen that the Dynkin diagram is O—O. The Dynkin diagram for SO(5) is O═● and for G_2 is O≡●.

Without filling in further details, we go to the final classification of all simple-root systems, which include the SU(n), SO(n), Sp(2n), and the exceptional algebras E_6, E_7, E_8, G_2 and F_4. For many of our concerns, the Dynkin diagram conveniently contains all the information needed, so we do not have to refer constantly to the commutation relations or to keep track of an explicit representation of the structure constants. If the Dynkin diagram has several pieces that are disconnected, then the algebra is semisimple, and each connected piece is simple.

Although the simple roots form a basis of root space, they do not form an orthonormal basis. The matrix that keeps track of the nonorthogonality is called the Cartan matrix. It has the elements,

$$A_{ij} = 2(\alpha_i, \alpha_j)/(\alpha_j, \alpha_j),$$

where the root-space vector α_i is the i-th simple root for the remainder of these lectures, and not a component of a root as it was before. The diagonal elements of the Cartan matrix are $A_{ii} = 2$. The matrix elements A_{ij} can be read off of the Dynkin diagram.

The simple roots form a basis of the Euclidean space into which root space is embedded, so each root or weight vector Λ in the weight space can be written as a linear combination of the simple roots, which we write in the convenient form,

$$\Lambda = \Sigma_i \, \lambda_i \, [2/(\alpha_i,\alpha_i)] \, \alpha_i \quad .$$

The longer simple roots are normalized conventionally to length-squared 2, so for algebras with all simple roots of the same length (SU(n), SO(2n), and E_n, called "simply-laced" Lie algebras), the factor $[2/(\alpha_i,\alpha_i)]$ is unity. The coefficients λ_i of Λ are called the dual coordinates. The Dynkin labels a_i of Λ (to which the λ_i are dual) are defined by

$$a_i = 2(\Lambda, \alpha_i)/(\alpha_i, \alpha_i) \quad ,$$

and are related to the λ_i by the Cartan matrix. This definition is of crucial importance.

The a_i are of great importance because of the following theorem: *for any root or for any weight of an irrep, the Dynkin labels a_i are integers.* For example, the Dynkin label for the weights of SU(2) is twice the magnetic quantum number, 2m.

Finally, we should discuss how to recover the physics: it is the quantum numbers that are measured in experiments, so it these that are needed, not the Dynkin labels. Fortunately, the solution to the problem of relating the quantum numbers to the Dynkin labels is extremely simple. The quantum numbers are linear combinations of the weight components, and so they correspond to projections of the weights onto various axes in weight space. Take the SU(3) example. In the Eightfold Way, the physically useful quantum numbers are I_3 and Y, or one of these and the electric charge, $Q = I_3 + Y/2$. The value of the charge is simply a projection of the weight onto and axis. Thus, we write,

$$Q(\Lambda) = \boldsymbol{\Sigma}_i \, q_i \, a_i \quad ,$$

where the numbers q_i are determined from the physics. They are the dual coordinates of the direction of the axis and do not depend on the representation, but only on the algebra.

The entire root system is the list of eigenvalues of the Cartan subalgebra when acting on the adjoint irrep, and the rules for working out the root diagrams are special examples of the rules needed for obtaining the Dynkin labels weight-by-weight for any irrep. It is convenient to discuss the full root system at the same time as the representation theory.

The representation theory of compact simple Lie algebras is summarized elegantly in terms of Dynkin labels a_i. We now discuss three problems and their solutions: (1) the enumeration of all finite-dimensional irreps of each algebra; (2) the weight system of each irrep; and (3) the computation of tensor (Kronecker) products of these irreps.

If there is an n-dimensional irrep **n** of G, then there exists an n-dimensional Hilbert space on which the generators of simple G act faithfully. The singlet **1** is also an irrep, but the generators all annihilate the singlet state, so the **1** is not a faithful irrep. The singlet is the only unfaithful irrep that we need here. (We often identify irreps by their dimensionality, an objectionable practice that is extremely convenient for low-lying irreps.) Each vector in the Hilbert space may be (partially) labeled by a weight. Recall that weight space is the lattice of all possible weights embedded in the rank(G) Euclidean space spanned by the simple roots. Perhaps some confusion will be avoided by stating that it should be clear from context whether "vector" refers to the Hilbert space vector in the representation or the vector in weight space being used to label the Hilbert space vector.

Weight space has a set of symmetries. If we take a weight $\mathbf{w}$ from the space and apply the transformation,

$$\mathbf{w}' = \mathbf{w} - \big(2(\mathbf{w},\boldsymbol{\alpha})/(\boldsymbol{\alpha},\boldsymbol{\alpha})\big)\,\boldsymbol{\alpha}\ ,$$

where $\boldsymbol{\alpha}$ is any root, then $\mathbf{w}'$ is also a vector in the weight space. This is called a Weyl reflection, and the set of Weyl reflections form a group. If all the weights of an irrep undergo a Weyl reflection, the new set of weights is equivalent to the old set. Thus, if the set $\{\mathbf{w}\}$ is a set of weights for a representation **r**, then $\{\mathbf{w}'\}$ is the set of weights for an equivalent representation. The set of weights related to $\mathbf{w}$ by all Weyl transformations is the Weyl orbit of **w**. The weights of an irrep can be decomposed into Weyl orbits.

We can use this symmetry to greatly simplify the problem of representation theory if we can find a convenient way of picking out just one weight on a Weyl orbit. In fact there is a simple solution to this problem: each distinct set of nonnegative Dynkin labels is a weight on a unique Weyl orbit, and every Weyl orbit has a representative in this set. This set of weights is called the positive Weyl chamber.

In a given irrep, some weights may be degenerate, that is, several vectors in Hilbert space may have the same weight, so that distinguishing them requires additional labels. However, there are always some weights in an irrep that are not degenerate, and one of those weights uniquely defines (up to a Weyl transformation) the irrep. That weight is called the highest weight. The highest weight has the property that it is annihilated by all the E_α, where α is a positive root. (This is the generalization to any simple algebra of the result in angular momentum theory, $J_+ |j,j\rangle = 0$.) Although there is more to prove, it may already be plausible that the following extremely important result is true: *Each and every finite dimensional unitary representation of a finite-dimensional Lie algebra is uniquely identified by the set of integers* $(a_1, \ldots, a_\ell)$, a_i *nonnegative. Each such set is a highest weight of one and only one irrep.*

The complete set of weights for each irrep can be derived from the highest weight and the Dynkin diagram. Once the weight system is in hand, the Dynkin labels can then be converted into the eigenvalues of a convenient set of diagonal generators.

The next problem is to list the remaining weights in the representation. There are two approaches: all the weights on the Weyl orbit of the highest weight are in the irrep, and they are nondegenerate weights. The remaining weights are on orbits "inside" of this orbit; one approach is to find the multiplicity of each of the inside orbits. This has been done for a huge number of representations of algebras up to rank 12 in a recent book by Bremmer, Moody and Patera [2], which the serious student will examine with enough care to be able to use it.

The second approach is more traditional. The lowering operators do not necessarily annihilate the vector of highest weight. In fact we could consider the infinite set of vectors formed by acting on the highest weight state $|h\rangle$ with all possible combinations of lowering operators but with no regard for the group structure. However, when the matrix elements are known, it is found that these series of ladder operator operation terminate, which leaves a finite number of vectors in the representation.

The following procedure tells when the matrix elements are zero. Suppose that the highest weight has components $\Lambda = (a_1, \ldots, a_\ell)$ in the Dynkin basis. Then the i-th simple root can be subtracted from Λ a_i times, which corresponds to the fact that the corresponding lowering operator can be applied to the highest weight a_i times. The $(a_i + 1)$-th application of the $E_{\alpha i}$ lowering operator gives zero, which is due to a zero in the matrix element. This rule can be used over and over until there are no new weights that have any positive Dynkin labels. This completes the construction of the representation, except for computing the degeneracy.

There is a general solution to the problem of finding the degeneracy of weight Λ' of representation with highest weight Λ. The solution was given by Freudenthal, although there are now variations of the Freudenthal formula that are easier to use [3]. The Freudenthal formula is a recursion relation that gives the degeneracy of a weight reached by subtracting n simple roots from the degeneracies of the weights where fewer simple roots are subtracted from the highest weight. The formula is complicated to use (even on a computer) because it fails to take advantage of the Weyl symmetries of the weight diagram. It is good to know that much more efficient formulas are available.

Tensor products of irreps of a simple Lie algebra are reducible into a direct sum of irreps. Thus, the product **R x R'** can be decomposed into a sum,

$$R \times R' = \Sigma_i R_i ,$$

where a given irrep may occur several times in the sum.

The tensor product can (in principle) be computed as follows: find each weight vector of **R** and **R'** and add them together in all combinations, which yields dim(**R**) x dim(**R'**) weights. Now find the highest weight and remove from the list all weights associated with the irrep with that highest weight. Then pick out of the remaining list the highest weight and remove the weights associated with the irrep with that highest weight, and so on, until no weights are left. This method is cumbersome, even on a computer, but is does illustrate the important point that the reduction of a tensor product can be done in weight space.

There are a number of aids for computing products of fairly low lying irreps by hand.

(1) Highest weights: The highest weight in $(a_1 \ldots a_\ell) \times (b_1 \ldots b_\ell)$ is $(a_1+b_1 \ldots\ldots a_\ell+b_\ell)$. There is also a rule for computing the irrep of second highest weight: subtract from the highest weight in the product the minimal sum of simple roots that connect together a nonzero component ai and a nonzero compont b_j . For example, if a_1 and b_1 are nonzero, simply subtract the Dynkin labels of the first root (the first row of the Cartan matrix) from the highest weight in the product. In the case of E_6 the product (100000)x(000001) (**27**x**78**) has highest weight (100001) and second highest weight (000100).

(2) Lowest weight rules: Add the lowest weight of **R'** to the highest weight of **R** and Weyl reflect this weight to the positive chamber. The resulting weight is the highest weight of the irrep of lowest weight in the decomposition of **R** X **R'**. The rule for finding the irrep of second lowest weight is analogous to the one above for finding the irrep of second highest weight. To the highest weight of the lowest weight irrep, add the same minimal chain as used for finding the irrep of second highest weight above, and then Weyl reflect that weight to the positive chamber.

As an example, consider the SO(7) product **48** x **35** = (101) x (002). The highest weight irrep is the **560** = (103). The second highest weight irrep is (103) - α_3 = (1 0 3) - (0 -1 2) = (1 1 1) = **512**. The lowest weight irrep has the weight (0 0 2) + (-1 0 -1) = (-1 0 1) on the same Weyl orbit as its highest weight. Weyl reflect with r_1, then r_2, then r_3 (r_i is the Weyl reflection with $\boldsymbol{\alpha}$ equal to the simple root α_i):

$$r_1(-1\ 0\ 1) = (-1\ 0\ 1) + (2\ -1\ 0) = (1\ -1\ 1);$$

$$r_2(1\ -1\ 1) = (1\ -1\ 1) + (-1\ 2\ -2) = (0\ 1\ -1);$$

$$r_3(0\ 1\ -1) = (0\ 1\ -1) + (0\ -1\ 2) = (0\ 0\ 1).$$

Thus, the lowest weight irrep in **48** X **35** is **8** = (0 0 1). The irrep of second lowest weight has on its Weyl orbit with the highest weight, (0 0 1) + (0 -1 2) = (0 -1 3). Reflecting with r_2 and then r_1, we find this irrep to be the **48** = (101), which actually occurs twice in **48** x **35**.

(3) Dimensionality and Index sum rules:

$$\mathrm{Dim}(\mathbf{R} \times \mathbf{R'}) = \mathrm{Dim}(\mathbf{R}) \times \mathrm{Dim}(\mathbf{R'}) = \boldsymbol{\Sigma}_i\ \mathrm{Dim}(\mathbf{R}_i) \quad ,$$

where Dim(**R**) is the dimensionality of **R**; and

$$\mathrm{Ind}(\mathbf{R} \times \mathbf{R'}) = \mathrm{Ind}(\mathbf{R})\cdot\mathrm{Dim}(\mathbf{R'}) + \mathrm{Dim}(\mathbf{R})\cdot\mathrm{Ind}(\mathbf{R'}) = \boldsymbol{\Sigma}_i\ \mathrm{Ind}(\mathbf{R}_i) \ ,$$

where the index Ind(**R**) is related to the Casimir invariant by the equation,

$$\mathrm{Ind}(\mathbf{R}) = [\mathrm{Dim}(\mathbf{R})/\mathrm{Dim}(G)]\ \mathrm{Casimir}(\mathbf{R}) \quad ,$$

and is related to the sum of the length-squared of all the weights of **R** by

$$\mathrm{Ind}(\mathbf{R})\ \mathrm{Rank}(G) = \boldsymbol{\Sigma}_\lambda\ (\lambda, \lambda) \quad .$$

Tables of dimensionalities and indices can be found many places.

(4) Crossing: If **R**x**R'** contains **R"**, then **R** x $\overline{\mathbf{R}}$**'** contains $\overline{\mathbf{R}}$**'**, where the overbar designates the conjugate irrep. For nonselfconjugate irreps, **R** and $\overline{\mathbf{R}}$ are inequivalent.

(5) **R** x $\overline{\mathbf{R}}$ always contains the singlet and adjoint representations. If **R** is self conjugate, then there are two cases: If the symmetric product contains the singlet, then the adjoint is in the antisymmetrized product and the representation matrices can then be brought to real form. All adjoint irreps are of this type. If the antisymmetrized product has the singlet, then the symmetrized product has the adjoint, and the matrices cannot be brought to real form; such irreps are called pseudoreal. The defining irrep of the symplectic groups is pseudoreal, and can be used to define a symplectic form. Recall the doublet of SU(2), which is isomorphic to Sp(2), is self conjugate and $(\mathbf{2}\times\mathbf{2})_A = \mathbf{1}$ and $(\mathbf{2}\times\mathbf{2})_S = \mathbf{3}$. There is no SU(2) transformation that brings the three Pauli matrices to real form.

(6) Congruency constraints: The weight space of G can in some cases be divided into classes of weights where no weight in one class can be reached by adding roots to a weight in another class. For example, no weight in a half-odd integer irrep of angular momentum can be reached from a weight of an integer spin representation by the action of the group generators. The case of SO(2n) is one of great concern in the next lecture. SO(2n) (n $\geq$ 2) has four congruency classes. Representative irreps of each class are, the adjoint, vector, and the two different spinors. All the irreps in a tensor product are in the same congruency class.

The congruency classes for SU(n) are identified by an integer in $\mathbb{Z}_n$ (n-ality), and the congruency of the irreps in the product of $r_p \times r_q$ with congruencies p and q, respectively, is p + q mod n. The congruency classes for SO(2n+1), Sp(2n) and E_7 are labelled by $\mathbb{Z}_2$. E_6 has three congruency classes labelled by the members of $\mathbb{Z}_3$, and G_2, F_4 and E_8 all have just one congruency class.

The four congruency classes of the SO(2n) algebras are slightly more complicated, since SO(4n) and SO(4n+2) differ. The congruency classes of SO(4n) are labeled by $\mathbb{Z}_2 \times \mathbb{Z}_2$. For example, the **1**, $\mathbf{8}_V$, $\mathbf{8}_S$, and $\mathbf{8}_C$ may be assigned to (0, 0), (1, 1), (1, 0) and (0, 1), respectively. (SO(8) has three distinct **8**'s due to the symmetry of its Dynkin diagram.) Thus, the irreps in $\mathbf{8}_V \times \mathbf{8}_S$ belong to the $(1,1) + (1,0) \approx (2,1) \approx (0,1)$ congruency class; this product contains the $\mathbf{8}_C$.

The four congruency classes for SO(4n+2) are labeled by $\mathbb{Z}_4$. For example, the **1**, **10**, **16**, $\overline{\mathbf{16}}$ of SO(10) are assigned to classes, 0,2,1,3, respectively. Thus, $\mathbf{16} \times \mathbf{16}$ and $\overline{\mathbf{16}} \times \overline{\mathbf{16}}$ have congruency 2, and the vector **10** appears in both of these products, but the adjoint cannot and does not appear in either product, since it has congruency 0.

References

1. For reviews of this approach, see, for example, B. G. Wybourne, Classical Groups for Physicists (Wiley, New York, 1974);
J. E. Humphreys, Introduction to Lie Algebras and Representation Theory (Springer, New York, 1972);
H. Georgi, Lie Algebras in Particle Physics (Benjamin/Cummings, Reading,1982);
R. Slansky, Physics Reports **79** (1981) 1.

2. M. R. Bremner, R. V. Moody and J. Patera, Tables of Dominant Weight Multiplicities for Representations of Simple Lie Algebras (Dekker, New York, 1984).

3. R. V. Moody and J. Patera, Bull. Am. Math. Soc, **7** (1982) 237.

COMPACTIFIED STRINGS AND THE HETERIOTIC THEORIES

Lecture 3

STRINGS AND THEIR COMPACTIFICATION FROM THE PARTICLE VIEWPOINT

R. Slansky

T-8, Theoretical Division
Los Alamos National Laboratory
Los Alamos, NM 87545 USA

The topics of this lecture are: (1) showing how infinite-dimensional Kac-Moody affine algebras can be spectrum generating algebras in (open) string theories, work I did with Louise Dolan [1]; (2) examining the spectrum of excited states of the heterotic string [2]; and (3) commenting on representations of affine algebras. We consider bosonic string models in D dimensions with d of the dimensions being a d-torus. The d-torus is a model for the compactified dimensions. After studying open strings, where the string formalism is quite simple, we look at the heterotic string, which is a closed string model with some formal similarities to the open string. The heterotic string appears to be consistent when interactions are included (the one loop amplitudes are finite and the integrands are modular invariant), and the gravitational and Yang-Mills anomalies cancel. "Part" of the heterotic string is a closed bosonic string in 26 dimensions and part is a closed superstring in 10 dimensions. We compute the spectrum for the $E_8 \times E_8$ version in 10 dimensions; the same methods can be used to analyze the $\mathrm{Spin}(32)/\mathbb{Z}_2$ model.

The slope of the Regge trajectory α' provides a scale of mass in string theories. Further mass scales R_I ($I = 1, \ldots, d$) appear to be required for each compactified dimension, although there may eventually be a physical principle that determines the ratios, R_I^2/α'. Each D-dimensional momentum eigenstate corresponds to an infinite number of (D-d)-dimensional states, just as in field theory. The mass operator of the compactified open bosonic string then has the form [3],

$$\alpha' M^2_{D-d} = N - 1 + \alpha' \Sigma_{I=1}{}^{d}\, p^I p^I,$$

where N is the number operator and the internal momentum components p^I are quantized in units of $1/R_I$, in analogy to compactification on a circle in Kaluza-Klein theories. The compactified dimensions are spacelike, so repeated upper-case latin indices imply a sum with metric tensor, δ_{IJ}. The number operator,

$$N = \Sigma_{n=1}{}^{\infty} \left(a_{-n}{}^{i} a_n{}^{i} + a_{-n}{}^{I} a_n{}^{I}\right),$$

is a sum of two pieces, one depending only on the modes for the internal space and the other depending only on the modes of the uncompactified space.

For arbitrary values of R_I, the states constructed from the $a_{-n}{}^{I}$ alone belong to a representation of a d-dimensional Heisenberg algebra (also called affine-$U(1)^d$) The full spectrum is a direct product of the representation of this affine $U(1)^d$ with the Fock spaces constructed from $a_{-n}{}^{i}$ and the zero mode operators. In the limit that $R_I^2 = \alpha'$, as described below, the $p^I p^I$ term in the mass becomes an integer and the degeneracy of states at each mass level above the tachyon increases. We construct the states corresponding to a type of compactification that requires $R_I^2 = \alpha'$ and $\sqrt{2\alpha'}p^I$ to take values on a weight lattice of a finite parameter, rank d, nonabelian Lie algebra G, restricted to those cases where the roots have equal lengths (the simply-laced algebras). The open string states become organized into representations of affine-G, where the nonabelian finite Lie subalgebra G of affine-G has rank d. The first excited level includes scalar massless particles in the adjoint representation of G; they are vectors for the closed string. The internal symmetry arises in a new way: it is neither an isometry of an internal space nor an explicit internal symmetry of the higher-dimensional theory. The fine tuning of R_I^2/α' to unity is encouraged by the appearance of this huge symmetry, although a better argument for setting R_I^2/α' to unity is lacking at present. Since string theories are S-Matrix theories in this formulation, G is also a symmetry of the states. G tends to be a rather large symmetry, but the formalism is not yet adequate for understanding how G is broken.

We approach the problem of finding the spectrum of the compactified string in two ways: first, we show by a simple counting argument, which will prove powerful in practical calculations of spectra, how the on-mass-shell physical states at each interger value of $\alpha' M^2_{D-d}$ are organized into representations of G; then, using a physical light-cone-gauge version of the Frenkel-Kac construction, we rewrite the physical operators in a form that explicitly displays the affine-G structure of the spectrum.

The (D-d)-dimensional mass operator is

$$M^2_{D-d} = 2p^+ p^- - \Sigma_{i=1}^{\infty} p^i p^i \quad ,$$

where the sum on i is over the noncompactified, transverse directions. It follows from the light-cone-gauge commutation relations that

$$[\alpha' M^2_{D-d} , a_n^i] = -n \, a_n^i \quad .$$

Note that a_n^i commutes with p^j so only $\exp(ip^I x^I)$ creates momentum in the transverse directions. It is easily confirmed that

$$\alpha' M^2_{D-d} |p\rangle = \left(N - 1 + \alpha' \Sigma_{I=1}^{\infty} p^I p^I \right) |p\rangle \, .$$

Since the Heisenberg algebras with internal indices commute with those with external labels, the problem separates into two pieces. The operators with noncompact-space labels only create Fock-space states of the (D-d-2)-dimensional Heisenberg subalgebra, where the states carry a (D-d)-dimensional label and the occupation-basis labels. We can then apply the internal-space operators to each of these states to obtain the complete set of physical states.

Since the a_n^I form a d-dimensional Heisenberg algebra, the number of independent operator combinations is given by the generating function (which is the inverse of the d-th power of the Euler function),

$$\Pi_{k=1}^{\infty} (1 - x^k)^{-d} = \Sigma_{n=0}^{\infty} M_n(d)\, x^n = 1 + dx + (d/2)(d+3)\, x^2 +$$

$$(d/6)(d+1)(d+8)\, x^3 + (d/24)(d+1)(d+3)(d+14)\, x^4 +$$

$$(d/120)(d+3)(d+6)(d^2 + 21\, d + 8)\, x^5 + \quad \ldots ,$$

where $M_n(d)$ is the degeneracy of the n-th state of the d-dimensional Heisenberg algebra representation. If we require the eigenvalues of $(\alpha' M^2_{D-d} + 1)$ to be nonnegative integers, then the last term in the mass is also a nonnegative integer; given a scheme to do this, the total degeneracy results from convoluting these numbers with the degeneracy.

The connection with G comes from solving this problem by imposing

$$R_I^2 = \alpha' \quad ,$$

and $\sqrt{2\alpha'}\, p^I$ to be any point on the infinite root lattice of a semisimple nonabelian finite algebra with roots of equal length. (If the roots of the algebra do not have equal lengths, then the particles in the irreps of G do not have equal (D-d)-dimensional masses.)

Let us first impose the condition that the components of the internal momentum are quantized to be

$$\alpha' p^I = (R/\sqrt{2})\, \Sigma_{J=1}^{\infty} N^J \alpha_J^{\ I},$$

where α_J is the J-th simple root vector, N^J are integers, and the $\sqrt{2}$

factor is inserted because of the customary normalization $(\alpha_I, \alpha_I) = 2$; all the radii are equal to R.

An easy way to find the torus that gives this quantization of the momentum is to recall that p^I generates translations in the I-th direction. Translation around the torus by some integer number of times will not change the state. Suppose that translating a point on the torus by $x_0^I = \sqrt{8}\pi R n^J \alpha^{*}{}_J{}^I$ always returns it to the starting point. (This defines a period on the torus.) Then the single-valuedness of the wave function on the torus is maintained if $x_0^I p_I = 2\pi(\text{integer})$. It follows that α^{*I} is a dual basis vector on weight space:

$$(\alpha^{*I}, \alpha_J) = \Sigma_{K=1}\, \alpha^{*}{}_K{}^I\, \alpha_K{}^J = \delta^{IJ} .$$

Finally, we can write this result as

$$\alpha' M^2{}_{D-d} = N + 1 + (1/2)(a, a)$$

for the open string, where the d-tuple of numbers a_I are the Dynkin labels for momentum. For algebras with roots of equal lengths, (a, a) is an even integer for any weight in the same congruency class as the adjoint.

As a more complicated example, the same construction can be done for SO(8); the first three levels are:

N = 0, **1**;

N = 1, **28**;

N = 2, $\mathbf{1} + \mathbf{28} + \mathbf{35}_v + \mathbf{35}_s + \mathbf{35}_c$;

N = 3, $5 \cdot \mathbf{1} + 3 \cdot \mathbf{28} + \mathbf{35}_v + \mathbf{35}_c + \mathbf{35}_s + \mathbf{350}$.

This construction makes it plausible that for special choices of the lattice of internal-momentum values, the string states are organized into representations of G.

It is possible to transform the operators a_n^I and $\exp(ip^I x^I)$ into a set of affine-G operators that generates the same Hilbert space. The construction is identical to the one given by Frenkel and Kac using the covariant vertex operator [4], although we work with operators that create physical states.

The Frenkel-Kac construction uses the moments of the vertex operator to rewrite the creation operators. Thus, consider the operators,

$$A_n(r) = \Sigma_I \; r^I a_n^I \quad ,$$

$$X_n(r) = (c_r/2\pi i) \int dz\, z^n :\exp(ir^I Q^I(z)/\sqrt{2}\,\alpha'): \quad ,$$

where the normal ordering means that a_n is to the right of a_{-n} for $n>0$ and p^I is to the right of x^I. The contour integral is taken in a counterclockwise sense about the origin of the complex z-plane. For the open string, $Q^I(z)$ is defined in terms of the light-cone operators,

$$Q^I(z) = x^I - 2i\alpha' p^I \ln(z) + i\sqrt{2\alpha'} \; \Sigma_{n\neq 0} \; (1/n)\, a_n^I z^{-n} \; .$$

where the roots r are normalized to $(r,r) = 2$. The cocycle factor c_r satisfies $c_r c_s = (-1)^{r.s} c_s c_r = \varepsilon(r,s)\, c_{r+s}$ $(r+s \neq 0)$, where $\varepsilon(r,s) = \pm 1$, and $c_r c_{-r} = 1$.

With these definitions, it is a standard (but lengthy) computation to obtain the algebraic equations satisfied by the operators $X_n(r)$ and $A_n(s)$:

$$[A_n(s), X_m(r)] = (s,r)\, X_{n+m}(r),$$

$$[X_n(r), X_m(s)] = \varepsilon(r,s)\, X_{n+m}(r+s) \text{ if } r+s \text{ is a root not zero,}$$

$$= 0 \qquad \text{if } r+s \text{ is not a root, but not zero,}$$

$$[X_n(r), X_m(-r)] = A_{n+m}(r) + n\delta_{n+m,0} \quad ,$$

$$[A_n(r), A_m(s)] = n\,(r,s)\,\delta_{n+m,0}.$$

The c_r factors are needed to make the brackets into commutators when $(r,s) = \pm 1$. This is the affine Kac-Moody Lie algebra of G where the roots of G have equal length. Of special importance are the c-number terms in the last two commutators; these terms represent the central extension for this representation. The physical states of the compactified bosonic string are in representations of the affine-G.

We turn now to the heterotic string. It is a "chiral" combination of a closed D = 26 bosonic string and a D = 10 superstring. The two chiralities in this instance are defined in the σ-τ space, where the solutions of the string wave equations with the closed string boundary conditions move either to the left (the solution is a fuction of $\tau + \sigma$ only), or to the right, (the solution is a function of $\tau - \sigma$). The heterotic string is constructed as a combination of a right-moving D = 10 superstring and a left-moving D = 26 bosonic string. Since each chiral sector is consistent by itself, the full theory with both left and right movers is also consistent.

The closed string solution was found in Euclidean space-time. In this section we assume some of the extra d dimensions are a d-torus. Thus, we have to reappraise the solution. In doing this, we *assume* that only the zero modes are affected by the change of geometry, and the solution for the nonzero modes does not have to be modified to guarantee that the string stays on the surface of the torus. The modification of the zero-mode solution is due to its more complicated topology; it is possible for a closed string to be wrapped around the torus like a rubber band around a newspaper, which has no analogue for Euclidean extra dimensions. In particular, the periodic boundary conditions allow the addition of a winding-number term,

$$X^I(\tau,\sigma) = x^I + 2\alpha' p^I \tau + 2Rw^I\sigma +$$

$$i\sqrt{\alpha'/2}\ \Sigma_{n\neq 0}(1/n)\left(a_n{}^I\, e^{-2in(\tau-\sigma)} + \tilde{a}_n{}^I\, e^{-2in(\tau+\sigma)}\right),$$

where

$$w^I = (1/\sqrt{2}) \, \Sigma_{J=1} \, n^J \, \alpha_J{}^I .$$

Thus, $\sigma \to \sigma + \pi$ is an identity transformation, if the winding number components n^J are integers and the basis vectors α_J $(J = 1, \ldots, d)$ for the extra dimensions are normalized to $\sqrt{2}$. (We have changed the notation from the previous discussion because this time we begin with the lattice rather than the momentum vectors, and so it is here that we fix the normalization to coincide with the root vectors. The normalization of the vectors $\alpha^*{}_J$ dual to the roots α_J depends on the lattice.)

Let us focus on the left-moving bosonic string. In the extra $26 - 10 = 16$ dimensions, the allowed solution to the bosonic wave equation is a function of $\sigma + \tau$. This constraint is imposed on the solution by adding to the action a Lagrange multiplier term of the form $\lambda_I[(\partial_\tau - \partial_\sigma)X^I]^2$. The string function $X^I(\sigma,\tau)$ is a function $X^I(\tau+\sigma)$ only if

$$\alpha' p^I = R\, w^I \qquad I = 1, \ldots, 16,$$

and all the $a_n{}^I$ $(n \neq 0)$ for the closed bosonic string equal to zero. This condition requires the winding number and the momentum to be on the *same* lattice. This condition is not obviously consistent with the quantum mechanics of the string, since wave-particle duality also restrict the momentum eigenvalues on the torus. But if we can find a solution, the closed bosonic string expansion for the extra 16 dimensions is

$$X^I(\sigma+\tau) = x^I + 2\alpha' p^I(\tau+\sigma) + i\sqrt{\alpha'/2}\, \Sigma_{n\neq 0} \, (1/n)\, \tilde{a}_n{}^I \, e^{-2in(\tau+\sigma)}.$$

The Lagrange multiplier term in the Lagrangian is crucial in finding the quantization rules. The $a_n{}^I$ satisfy the usual commutation relation, since there are no constraints on the nonzero modes. However, the zero-mode operators satisfy a *modified* form of the $[x, p]$ commutation relations,

$$[x^I , p^J] = (1/2)\ i\ \delta^{IJ} \quad ,$$

where the (1/2) factor arises because of the constraint. It is required for the consistency of the theory.

Let us return to the constraint on the zero mode. For fixed τ, the value of X^I increases by $2\pi R w^I$ as σ is increased from 0 to π; the point $X^I(\tau)$ must be identically the same point as $X^I(\tau+\pi)$. Thus, the set of coordinates $X^I + \sqrt{2} R\pi n^J \alpha_J{}^I$ for all integers n^J in the 16-dimensional space of extra dimensions all identify the same point. (This is a rather fancy definition of a torus: a torus is a Euclidean space mod a lattice defined on the space.) The allowed momentum eigenvalues on this lattice can be calculated as before—almost. The difference is the commutation relation of x and p, which is the statement of wave-particle duality on this lattice. The constraint is that $p^I\ x_0{}^I$ is π times an integer (not 2π), where $x_0{}^I = (\pi R\sqrt{2})\ \Sigma_{J=1}\ n^J\ \alpha_J{}^I$. Thus, the momentum resulting from this quantization condition is

$$p^I \;=\; [1/(R\sqrt{2}\,)]\;\; \Sigma_{K=1}\ N^K\ \alpha^*{}_K{}^I \;,$$

where α^{*I} is again dual to α_J. We must now require that $\alpha' p^I = R\,w^I$ so the heterotic closed string will be a function of $\sigma + \tau$. Combining the above results, we arrive at the constraint (for $\alpha' = R^2$),

$$\Sigma_{J=1}\ N^J\ \alpha^*{}_J{}^I \;\;=\;\; \Sigma_{J=1}\ n^J\ \alpha_J{}^I\ .$$

This condition states that the lattice of momentum values must exactly coincide with the lattice that defines the torus of the extra 16 dimensions. Thus, the lattice is self dual. Just as for the open string (where we did not have a self duality constraint), the basis vectors of the lattice must have all roots of the same length, chosen to be length-squared 2 by convention. The lattice is then called an even self dual lattice.

There are not many even self-dual Euclidean lattices. There is one in 8 dimensions, which is defined by the root lattice of E_8. The next ones occur in 16 dimensions, which is fortunate, or the compactification of the left-movers only would not be consistent with the quantization conditions. One is a root lattice where the basis of the lattice are the simple roots of the algebra of $E_8 \times E_8$. The other self-dual lattice is a weight lattice. It is the weight lattice of SO(32) where the zero-congruency weights (0, 0) plus the weights congruent to just one spinor (1, 0) are included. (This lattice is called Spin(32)/$\mathbb{Z}_2$ to signify that you may start with the full SO(32) weight space, including all four congruency classes, and then throw out two of the congruency classes, which are chosen to be the vector set (1, 1) and one of the spinor sets (0, 1).) Thus, the gauge groups for the heterotic string are greatly restricted. However, this restriction is consistent with the restriction to string theories where the gravitation and Yang-Mills anomalies cancel [5], so in fact, if we pay attention to the restrictions from the zero-slope quantum field theory limits, the restriction to self-dual lattices is no restriction at all. Moreover, in so far as open superstring theories do have anomalies, the analysis of open string theories cannot lead to a consistent theory. However, we will find that the analyses of the heterotic strings are similar to those of open strings, except that the groups are limited to $E_8 \times E_8$ and SO(32).

We now proceed, assuming that the extra 16 dimensions are compactified on a even self-dual lattice. The 10-dimensional right-moving superstring must be combined with the left moving bosonic string. The mass operator in 10 dimensions is

$$\alpha' M_{10}{}^2 = 2N + 2\tilde{N} - 2 + 2\alpha' \Sigma_{I=1} p^I p^I \quad ,$$

where the number operator for the right movers N includes both bosonic and fermionic operators, and $\tilde{N}$ for the left movers includes bosonic operators for both 10-dimensional space time and for the 16 extra dimensions. There is one further constraint on the closed string. Since a shift in the origin of the parameter σ does not change the string, the physical states must carry zero eigenvalue of the generator of the shift, or

$$N = \tilde{N} - 1 + \alpha' \sum_{I=1} p^I p^I \ .$$

Just as for the open string, the 10-dimensional space-time quantum numbers can be factored from the compactified directions, so the mass operator can be written,

$$\alpha' M_{10}^{\ 2} = 4\tilde{N} \ ,$$

where $\tilde{N}$ is a sum of a 10-dimensional piece and an internal 16-dimensional piece. Thus, the mass cannot be negative, since the smallest eigenvalue of $\tilde{N}$ is 0.

Having established that the p^I correspond to weights on a lattice, we can now evaluate $\alpha' \sum_I p^I p^I = (1/2)\,(a, a)$, where a denotes the list of Dynkin indices for the weight. In the $E_8 \times E_8$ case, all possible weights are allowed because there is only one congruency class. Let us now evaluate $\alpha' \sum p^I p^I$ to find the $E_8 \times E_8$ spectrum of the heterotic string levels. The ground state of the left movers is $|0\rangle$ with $\tilde{N} = 0$, and the right moving superground state is $|i\rangle$ plus $|a\rangle$, where the N value of the superground state is 0. (Although this is a closed string, there is only one super set of $a_n^{\ i}$ and $S_n^{\ a}$. The constraints on this closed string break the expected $N=2$, $D=10$ supersymmetry to $N=1$, $D=10$ supersymmetry.) The ground state is not merely the direct product of these two ground states, since that state violates the constraint of σ translation invariance. The lowest mass states the direct product of $|i\rangle + |a\rangle$ with the states $\tilde{a}_{-1}^{\ I}|0\rangle$ and $|a^2{=}2\rangle$, which form an $N=1$ supermultiplet in the adjoint of $E_8 \times E_8$ (a super $E_8 \times E_8$ Yang-Mills multiplet), and with $\tilde{a}_{-1}^{\ j}|0\rangle$, which forms the $N=1$, $D=10$ supergravity multiplet. There are 480 states of the form $|(a, a) = 2\rangle$ and the 480 weights are the 480 nonzero roots of $E_8 \times E_8$.

The calculation of the next levels is greatly simplified if a table of orbit decompositions of E_8 irreps is available, along with the weight lengths. In the following table, we denote the orbit by the weight in the positive chamber in [...], and irreps by the highest weight (...).

Irrep	(a,a)	1	[00000010]	[10000000]	[00000100]	E_8 irrep
(00000010)	2	8	1			**248**
(10000000)	4	35	7	1		**3875**
(00000100)	6	140	35	7	1	**30380**
(00000020)	8	120	29	6	1	**27000***
(00000001)	8	370	111	29	6	**147250***

There is one orbit, the one with the highest weight, missing from this table for the **27000** and the **147250**. The irreps with the next longest highest weight is the (10000010), with (a, a) = 10. The numbers under the orbits gives the degeneracy of the orbit in the irrep listed on the left.

The $E_8 \times E_8$ irreps for the first massive level is: A (**1**,**1**) of $E_8 \times E_8$ in the **44**X(**44** + **84** + **128**) of SO(9); a (**248**,**1**) + (**1**,**248**) with spin **9**X(**44** +**84** + **128**); and (**3875**,**1**) + (**1**,**3875**) + (**248**,**248**) + 2(**1**,**1**) with spin **44**+**84**+**128**. Note that this last set makes it clear that the heterotic string has an $E_8 \times E_8$ symmetry of states only, and not the full affine symmetry. The 10-dimensional Poincare' algebra does not commute with the full affine $E_8 \times E_8$, but just the finite $E_8 \times E_8$.

Much is known about the irreps of the affine algebras. The irreps of greatest interest to physicists are those with a highest weight. We have found that one representation can be constructed from Fock-space operators $a_{-n}{}^{I}$ and a set of zero-mode operators $\exp(ix^I p^I)$ where p^I is a vector on the weight lattice. These operators can be written as contour integrals of vertex operators, which are the generators for the basic representation of affine-G. Other representations are obtained if the c-number terms (the central terms) have other coefficients taken from a restricted set of values.

Other representations of highest weight can be designated by their highest weight, just as for finite algebras. What we have constructed so far is the representation that begins with the singlet of G. However, there are irreps affine-G that begin with any finite-dimensional irrep of G. We will give a few examples. It may be that knowledge of affine irreps of highest weight may eventually become nearly as important in mathematics and theoretical physics as their finite G counterparts already are. Finally, we give an example how other representations of the affine algebras may be obtained with different sets of quantum operators.

Let us review the states of the basic irrep of affine-SU(2). The highest weight is a singlet. The SU(2) weight lattice is an even lattice, and so the zero mode states states of the Fock space construction are labeled by an integer n, where the mass of $|I, n\rangle$ is $n^2 + |I|$. For affine-SU(2), there is only one Heisenberg algebra. Now we can construct the states out of prducts of a_{-n}'s and exp(inx) acting on $|0\rangle$.

The reason for repeating these results is to show that there is an equivalent irrep that begins with the doublet of SU(2). This can be seen from the Dynkin diagram for the affine-SU(2) algebra. The Dynkin diagram from these infinite dimensional Lie algebras is the extended diagram of its finite counterpart. Thus, for affine-SU(2), the diagram is O====O. (Four lines means 180°.) If the first dot corresponds to the extended dot, the the Dynkin highest weight designation is (1, 0). However, it would seem that we should get an equivalent irrep from (0, 1) since it would be obtained from a symmetry operation on the Dynkin diagram. In fact, it is. Then the weights at each level fall into 1/2 odd integer irreps of SU(2). In the same way, we know that there are 4 equivalent affine-SO(8) irreps starting respectively with **1**, $\mathbf{8_v}$, $\mathbf{8_s}$, and $\mathbf{8_c}$. Similarly, affine E_6 has three basic irreps, beginning with **1**, **27** and $\overline{\mathbf{27}}$.

The SU(2) irreps of each of these affine-SU(2) irreps are related by the supersymmetry of OSp(1/2). Thus, one can expect some of the highest-weight irreps of affine superalgebras to be quite simple.

The construction of irreps not equivalent to the basic irrep is much more complicated. The solution is known in the mathematics literature.

There are partition functions for each of these irreps. These formulas are extensions of the partition function, which is the generating function for the degeneracy of states at each level of a d-dimensional Heisenberg algebra. Since this generating function is identical to the degeneracy of states of boson creation and annihilation operators, we can find (because of this mathematical result) the irrep of the Heisenberg algebra from the boson operators. Thus, if we can find quantum operators that give a partition function that is equal to the generating function of some other affine irrep, then we can construct a quantum theory based on that irrep. Let us look at a simple example, the (0, 2) irrep of affine-SU(2).

Consider the set of Neveu-Schwarz operators,

$$\{b_r^{\,i}, b_s^{\,j}\} = \delta^{ij}\,\delta_{r+s,0} \quad ,$$

where r and s take on 1/2 odd integral values, and i and j transform as a irrep of some algebra, to be chosen. These operators act on a highest weight vector $|0\rangle$. Just as in the Neveu-Schwarz model, the states can be divided into even and odd "G-parity" sectors. (The language comes from the dual pion model, which is a bosonic string model made out of anticommuting operators. These operators were combined with the Ramond operators in the discovering of the superstring theories. The formalism was latter straightened out by Green and Schwarz, by transforming to new operators that had 10-dimensional spinor indices.)

Let us select the i index of b to be the vector 3 of SO(3), which is also the adjoint of SO(3). We now look at the odd-G-parity sector of the Fock space created by these operators. The first few levels are $b_{-1/2}^{\,i}|0\rangle$, $b_{-3/2}^{\,i}|0\rangle$, $b_{-1/2}^{\,i}\, b_{-1/2}^{\,j}\, b_{-1/2}^{\,k}|0\rangle$, and so on, and the zeroth level is **3**, the first level is **3**+**1**, the second level is **1**+**3**+**3**+**5**, etc.

References

1. L. Dolan and R. Slansky, Phys. Rev. Lett. **54** (1985) 2075.

2. D. Gross, J. Harvey, E. Martinec and R. Rohm, Phys. Rev. Lett. **54** (1985) 502, and Nucl. Phys. B**256** (1985) 253.

3. E. Cremmer and J. Scherk, Nucl. Phys. B**103** (1976), 399.

4. I. B. Frenkel and V. G. Kac, Invent. Math. **62** (1980) 23;
J. Lepowsky and R. L. Wilson, Commun. Math. Phys. **62** (1978) 43;
G. Segal, Commun. Math. Phys. **80** (1981) 301;
P. Goddard and D. Olive, in Vertex Operators in Mathematics and Physics, ed. J. Lepowsky et al (Springer, New York, 1984), p. 419.

5. M. B. Green and J. H. Schwarz, Phys. Lett. **149B** (1985) 117.

TOROIDAL COMPACTIFICATION OF CLOSED BOSONIC STRINGS

Lecture 4

STRINGS AND THEIR COMPACTIFICATION FROM THE PARTICLE VIEWPOINT

R. Slansky

T-8, Theoretical Division
Los Alamos National Laboratory
Los Alamos, NM 87545 USA

The goal of this lecture is to show something of the algebraic structure of compactified closed bosonic strings. The works reported here was done in collaboration with Stuart Raby [1] and the recent work on Lorentzian lattices includes Louise Dolan. We study the boundary conditions of the first-quantized compactified closed bosonic string. Demanding that internal symmetries result from the Frenkel-Kac construction, we show at the tree level that the "left" and "right" lattices must be self-dual if all masses are integers; they can be chosen independently to be $E_8 \times E_8$ or $Spin(32)/\mathbb{Z}_2$ for 16 compactified dimensions. It is also shown how to get a group $G \times G$ on a slightly different lattice, which gives tachyons of two different mass values, including tachyons in nontrivial representations of $G \times G$. We discuss here the closed bosonic string; it would be interesting to apply some of these considerations to the heterotic string.

The non-interacting closed bosonic string in light-cone gauge satisfies a two-dimensional free wave equation. The solution is

$$X^i(\tau,\sigma) = x^i + 2\alpha' p^i \tau + 2N^i R\sigma + i\sqrt{\alpha'/2}\,\Sigma_{n\neq 0}(1/n)\left(a_n{}^i\, e^{-2in(\tau-\sigma)} + \tilde{a}_n{}^i\, e^{-2in(\tau+\sigma)}\right) \quad , \qquad (1)$$

where the coefficients are chosen so that in the first quantized theory, the corresponding operators satisfy canonical commutation relations, e.g.,

$$[x^i, p^j] = i\delta^{ij} \qquad (2)$$

The indices i, j, ... run from 1 to 24 and label the directions transverse to the light-cone components, $X^+(\tau, \sigma) = x^+ + 2\alpha' p^+ \tau$ and the dependent operator $X^-(\tau, \sigma)$ is gotten from the constraint equations. The term $2N^I R\sigma$ in (1) is a "winding number" term and can be nonzero for the compactified dimensions. As the parameter σ goes from 0 to π, the line traced out by $X^I(\tau, \sigma)$ is a closed curve, so $X^I(\tau, \pi)$ and $X^I(\tau, 0)$ must designate the same point on the compactified space. (I, J, K label the components of $X^i(\sigma, \tau)$ in the compactified dimensions.) In principle the "radii" R^I may differ for each direction, but the generality of our arguments are not affected by setting $R^I = R$.

In the first quantized theory N^I and $\sqrt{\alpha'}\, p^I$ (or $\sqrt{2\alpha'}\, p^I$) become operators with integer eigenvalues, which, however, must satisfy the constraints on the system. As is well known, it simplifies the problem to make a canonical transformation of these zero-mode operators to a set where the constraints are diagonal. The $X^i(\tau, \sigma)$ can be separated into a "left-moving" piece (in the τ-σ space), which is a function of $\xi_L = \tau + \sigma$, and a "right-moving" piece, which is a function of $\xi_R = \tau - \sigma$ only:

$$X^i(\tau, \sigma) = X_L{}^i(\xi_L) + X_R{}^i(\xi_R) \quad , \qquad (3)$$

where

$$X_L{}^i(\xi_L) = x_L{}^i + \alpha' p_L{}^i \xi_L + i\sqrt{\alpha'/2}\ \Sigma_{n\neq 0}(1/n)\, \tilde{a}_n{}^i\, e^{-2in\xi_L} \quad , \qquad (4)$$

and an identical equation for $X_R{}^i(\xi_R)$, except that L is replaced by R and $\tilde{a}_n{}^i$ by $a_n{}^i$. The canonical commutation relations of the zero-mode operators in the compactified directions are

$$[\, x_L{}^I , p_L{}^J] = i\delta^{IJ} \quad , \qquad [x_R{}^I , p_R{}^J] = i\delta^{IJ} \quad , \qquad (5)$$

and the remaining commutators are zero. Note that we selected a class of theories with two sets of zero-mode operators for the compactified dimensions. The classical heterotic strings have only one set of zero-mode operators for the compactified dimensions. The canonical transformation between this set of operators and the zero-mode operators in (1) is

$$x^I = x_L{}^I + x_R{}^I \quad , \qquad 2p^I = p_L{}^I + p_R{}^I \quad ,$$

$$2R\, N^I = \alpha'(p_L{}^I - p_R{}^I) \quad , \qquad \alpha'\tilde{x}^I = R(x_L{}^I - x_R{}^I) \tag{6}$$

Thus, it is trivially confirmed that $[\tilde{x}^I, N^J] = i\delta^{IJ}$, etc, so $\tilde{x}^I$ is conjugate to N^I, although $\tilde{x}^I$ does not appear in (1). Note that the winding number of a string can be transformed, for example, by self interactions, so it is an independent quantum operator unless further constraints are imposed on the system.

The constraint equations are

$$(\acute{X} \pm X')^2 = 0, \tag{7}$$

which are also written as

$$(\acute{X}_L)^2 = (\acute{X}_R)^2 = 0 \;. \tag{8}$$

The zero-mode projections of (8) define the dependent operators $p_L{}^-$ and $p_R{}^-$. This projection of (8) may be written as

$$(1/4)\,\alpha'\,M_L{}^2 = (1/2)\,\mathbf{w}_L{}^2 + \tilde{N} - 1 \quad ,$$

$$(1/4)\,\alpha'\,M_R{}^2 = (1/2)\,\mathbf{w}_R{}^2 + N - 1 \quad ,$$

$$\mathbf{w}_{L,R} = \sqrt{\alpha'/2}\;\; \mathbf{p}_{L,R} \;, \tag{9}$$

where the number operators N and $\tilde{N}$ are

$$N = \Sigma_{n=1}^{\infty} a_{-n}^{i} a_{n}^{i} \quad , \quad \tilde{N} = \Sigma_{n=1}^{\infty} \tilde{a}_{-n}^{i} \tilde{a}_{n}^{i} \quad , \tag{10}$$

and $\mathbf{w}_{L,R}$ are vectors in the compactified directions. The invariance of the string under a redefinition of the origin in σ imposes that $M_L = M_R$, and the total mass operator is

$$2M^2 = M_L^2 + M_R^2 = 2M_L^2. \tag{11}$$

We require that the compactified theory has large internal symmetries that can be found from the Frenkel-Kac construction. Then $\mathbf{w}_L$ and $\mathbf{w}_R$ lie on (independent) even lattices with shortest nonzero vectors corresponding to the roots of a semisimple simply laced Lie algebra. (An even lattice is one for which every vector has length-squared equal to an even integer.) A careful adjustment of the ratio of α' to the scale of compactification is required so that the smallest mass interval due to the internal momenta is equal to that of the number operators. We consider this case first, then the case where $(1/2)\,\alpha' M_{L,R}^2$ may have fractional eigenvalues, and $\mathbf{w}_L$ and $\mathbf{w}_R$ are on weight lattices

We first impose periodicity on the compactified space. This can be done by noting that p_L^I (and p_R^I) are translation operators in the $\mathbf{x}_L$ (and $\mathbf{x}_R$) spaces. These spaces are toruses, which can be viewed as Euclidean spaces mod the units of translation,

$$\mathbf{x}_L^{0} = \sqrt{2}\; r_L\, \pi\, \boldsymbol{\zeta}_L \quad , \qquad \mathbf{x}_R^{0} = \sqrt{2}\, r_R\, \pi\, \boldsymbol{\zeta}_R, \tag{12}$$

where by convention, the vectors $\boldsymbol{\zeta}_L$ (and $\boldsymbol{\zeta}_R$) are on even lattices, so the magnitudes of the vectors $\mathbf{x}_L^{0}$ are $2\pi r_L$ times integers. (*A priori*, the vectors of length $\sqrt{2}$ in $\boldsymbol{\zeta}_L$ and $\boldsymbol{\zeta}_R$ are not required to be roots of a Lie algebra.) All points in the $\mathbf{x}_L$ Euclidean space that differ by $\mathbf{x}_L^{0}$ are coordinates for the same point on the torus. (Equation (12) and others like it that involve the vectors on the lattices should be read as "the set of vectors on the left hand side is equal to the set of vectors on the right hand side.") A translation by $\mathbf{x}_L^{0}$ cannot change a physical state, so

$$e^{ip_L \cdot x_L^0} \,|\text{physical}\rangle = e^{2\pi i(\text{integer})}\,|\text{physical}\rangle \quad . \tag{13}$$

Thus, the eigenvalues of $\mathbf{p}_L$ must be in the set,

$$\mathbf{p}_L = (\sqrt{2}/r_L)\ \boldsymbol{\zeta}_L^* \quad , \qquad \mathbf{p}_R = (\sqrt{2}/r_R)\ \boldsymbol{\zeta}_R^* \quad , \tag{14}$$

where $\boldsymbol{\zeta}_L^*$ is a lattice dual to $\boldsymbol{\zeta}_L$. Equations (9) and (14) yield that

$$\mathbf{w}_L = (\sqrt{\alpha'}/r_L)\ \boldsymbol{\zeta}_L^* \quad , \qquad \mathbf{w}_R = (\sqrt{\alpha'}/r_R)\ \boldsymbol{\zeta}_R^* \tag{15}$$

must be even lattices. This requires adjusting the ratios, r_L^2/α' and r_R^2/α', since $\boldsymbol{\zeta}_L$ and $\boldsymbol{\zeta}_R$ were chosen to be even. Substituting (15) into (12), we also have

$$\mathbf{x}_L^0 = \sqrt{2\alpha'}\ \pi\ \mathbf{w}_L^* \quad , \quad \mathbf{x}_R^0 = \sqrt{2\alpha'}\ \pi\ \mathbf{w}_R^* \quad . \tag{16}$$

The closed-string boundary conditions that are imposed on (1) may be viewed in two ways. First, the "winding-number lattice" **N** is defined by the canonical transformation (6). Thus, $\sqrt{2}\ \mathbf{N}$ is the even lattice,

$$\sqrt{2}\ \mathbf{N} = (\sqrt{\alpha'}/R)\ (\mathbf{w}_L - \mathbf{w}_R) = (\sqrt{\alpha'}/R)\ \boldsymbol{\xi} \quad , \tag{17}$$

which also restricts the ratio α'/R^2. Thus, the relative orientation of lattices $\mathbf{w}_L$ and $\mathbf{w}_R$ must be such that $\boldsymbol{\xi} = \mathbf{w}_L \pm \mathbf{w}_R$ is indeed a lattice. (The set $\mathbf{w}_L - \mathbf{w}_R$ coincides with the set, $\mathbf{w}_L + \mathbf{w}_R$.) Second, as σ goes from 0 to π at fixed τ, $X^I(\sigma, \tau)$ must return to the same point on the torus. Thus, $2\pi R\ \mathbf{N}$ must be a vector on the lattices (16). From (1) and (16), this boundary condition requires that

$$\sqrt{2}\ \mathbf{N} = (\sqrt{\alpha'}/R)\ (\mathbf{w}_L^* - \mathbf{w}_R^*) \quad . \tag{18}$$

Thus, the boundary conditions can be satisfied only if the sum of the left and right lattices is identical to the sum of their duals. For example, if $w_L = w_R$ is an even self-dual lattice such that $\xi = w_L$, then

$$\alpha' = r_L^2 = r_R^2 = R^2 \quad . \tag{19}$$

The choice of even lattices for w_L and w_R is restricted to E_8 for 8 compactified dimensions, $E_8 \times E_8$ or $Spin(32)/\mathbb{Z}_2$ for 16 dimensions. The next even self-dual lattices appear in 24 dimensions.

Remaining solutions must satisfy the self-duality constraints, but ξ need not necessarily be even. However, not all self-dual lattices lead to solutions of the closed string. For example, in 12 dimensions there is an odd self-dual lattice composed of the weights in the adjoint and vector congruency classes of SO(24). Here the Frenkel-Kac construction fails because there is no way to make the shortest weights of w_L and w_R into the roots of rank 12 simply laced algebras.

We now turn to the case where w_L is an $E_8 \times E_8$ lattice and w_R is a $Spin(32)/\mathbb{Z}_2$ lattice. This combination satisfies (17) and (18), if ξ is indeed a lattice. The basis vectors of both w_L and w_R can be written in an SO(16) X SO(16) basis without distorting either root lattice. Thus, the roots of $E_8 \times E_8$ are in the **(120,1)** + **(1,120)** + **(128,1)** + **(1,128)** of SO(16) X SO(16), where **120** is the adjoint and **128** is one of the spinors of SO(16). Similarly for $Spin(32)/\mathbb{Z}_2$, the lowest lying representations with weights on the lattice include the adjoint **496** = **(120,1)** + **(1,120)** + **(16,16)** and one of the spinors **32768** = **(128,128)** + **(128',128')**, where **128'** is the other SO(16) spinor. We now add vectors of w_L to w_R; all that is needed are the congruency classes of the final set.

SO(16) has 4 congruency classes, which we label a (adjoint), s (spinor), s' (other spinor) and v (vector). The addition rules for the weights of one SO(16) are those of $\mathbb{Z}_2 \times \mathbb{Z}_2$, where a $\approx$ (0, 0), s $\approx$ (1, 0), s' $\approx$ (0, 1) and v $\approx$ (1, 1). Thus, the congruency classes of the $E_8 \times E_8$ lattice are (a, a) + (a, s) + (s, a) + (s, s), which closes on itself under addition. (This proves that the sum of two $E_8 \times E_8$ lattices can be arranged to be even, obviously.) The SO(16) X SO(16) congruency classes of the Spin(32)/$\mathbb{Z}_2$ lattice are (a, a) + (s, s) + (s', s') + (v, v), which again closes under addition. The sum of an $E_8 \times E_8$ and Spin(32)/$\mathbb{Z}_2$ lattices falls into classes,

$$\xi = (a,a) + (s,s) + (s',s') + (v,v) + (s,a) + (a,s) + (s',v) + (v,s'). \quad (20)$$

The length squared of all spinor and adjoint weights are even integers. However, all vector weights are odd, so the lattice vectors in the last two classes of (20) are odd. Thus, ξ has vectors of length-squared 2, 3, 4...

We now check the consistency requirement that **p** and **N** are the translation operators of the **x**, and $\tilde{\mathbf{x}}$ spaces, respectively. If $\mathbf{x}^0$ and $\tilde{\mathbf{x}}^0$ are the lattice units, then

$$e^{i(\mathbf{p}.\mathbf{x}^0 + \mathbf{N}.\tilde{\mathbf{x}}^0)}|\text{physical}\rangle = e^{2\pi i(\text{integer})}|\text{physical}\rangle . \quad (21)$$

From (6), all eigenvalues of $\mathbf{p}.\mathbf{x}^0$ are $\pi\xi.\xi$, where the scalar product is taken for any two vectors from ξ. The unit of lattice in $\tilde{\mathbf{x}}^0$ is

$$\tilde{\mathbf{x}}^0 = \pi R \sqrt{2/\alpha'}\ \xi , \quad (22)$$

so the eigenvalues of $\mathbf{N}.\tilde{\mathbf{x}}^0$ are also $\pi\xi.\xi$. If $\mathbf{w}_L$ and $\mathbf{w}_R$ are both $E_8 \times E_8$ lattices. Then ξ is the even $E_8 \times E_8$ (or Spin(32)/$\mathbb{Z}_2$) lattice, so $\mathbf{p}.\mathbf{x}^0$ and $\mathbf{N}.\tilde{\mathbf{x}}^0$ each have eigenvalues 2π(integer).

If ξ is an integer lattice, then $\alpha' = 2R^2$. In this case $\mathbf{p}.\mathbf{x}^0$ and $\mathbf{N}.\tilde{\mathbf{x}}^0$ can be odd integers, so we must show that the constraints expressed in (17) and (18) impose that when one is odd, the other must be also. For $\mathbf{w}_L = E_8 \times E_8$, and $\mathbf{w}_R$ = Spin(32)/$\mathbb{Z}_2$, if $\mathbf{p}.\mathbf{x}^0$ is an odd integer, then so is $\mathbf{N}.\tilde{\mathbf{x}}^0$, since they will both be in (s', a) or (a, s') of (20). This confirms the physical consistency of the solution.

We now give a preliminary discussion of the G X G case [3]. Suppose we place $\mathbf{w}_L$ and $\mathbf{w}_R$ on weight lattices. The the analysis of

(13) is slightly modified. Further, suppose $r_L = r_R = R = \sqrt{\alpha'}$. The boundary condition that the string is closed can be written,

$$\zeta_L - \zeta_R = \sqrt{2}\, N \quad , \tag{21}$$

where we have changed $\zeta_R \longrightarrow -\zeta_R$ as a set of weight vectors.

From the canonical transformation (6), we also have

$$w_L - w_R = \sqrt{2}\, N \ . \tag{22}$$

Finally, $w_{L,R}$ and $\zeta_{L,R}$ are related by a condition similar to (13),

$$\exp(ix_L{}^0.p_L{}^{op} - ix_R{}^0.p_R{}^{op})| x_L, x_R\rangle = \exp(2\pi i \text{ integer})| x_L, x_R\rangle, \tag{23}$$

which implies that

$$\zeta_L.w_L - \zeta_R.w_R = \text{integer} \quad . \tag{24}$$

Thus, the lattice $\zeta = \{\zeta_L , \zeta_R\}$ is dual to the lattice $w = \{w_L, w_R\}$. The constraints implied by (21) and (22) can then be solved if $\zeta = w^*$ is a self dual Lorentzian lattice with the metric given by

$$w_1 \cdot w_2 = w_{1L} \cdot w_{2L} - w_{1R} \cdot w_{2R} \quad . \tag{25}$$

Thus, although w_L by itself is not self dual, it is possible to make an even self dual lattice by doubling the space $w = \{w_L, w_R\}$ and imposing a Lorentzian metric.

Let us study a nontrivial example how this works. Consider the weight lattice of SU(3) X SU(3), and let a_1 and a_2 be the basis vectors, which correspond to the highest weights of the $\mathbf{3}$ and $\overline{\mathbf{3}}$ representations, respectively. Then we can form a basis of this 4-dimensional lattice with the vectors,

$$v_1 = (a_1\ ,\ a_2 - a_1)\quad ,\quad v_2 = (a_2,\ a_1 - a_2)$$

$$v_3 = (a_1 - a_2\ ,\ -a_1)\quad ,\quad v_4 = (a_1 - a_2\ ,\ a_2) \tag{25}$$

Note that these vectors are all null vectors. They are also self dual, i.e., there are a set of vectors v_i^* satisfying

$$v_i^* \, v_j \;=\; \delta_{ij} \tag{26}$$

It is easily confirmed that the dual vectors are $v_1^* = v_2$, $v_2^* = v_1$, $v_3^* = v_4$, and $v_4^* = v_3$.

This is an even lattice under the Lorentzian scalar product. The most general vector is

$$v \;=\; \begin{vmatrix} (n_1+n_3+n_4)\ a_1 \ + (n_2-n_3-n_4)\ a_2 \\ (-n_1+n_2-n_3)\ a_1 + (n_1-n_2+n_4)\ a_2 \end{vmatrix}\ , \tag{27}$$

where the metric for the scalar product $a_i.a_j$ is

$$\frac{1}{3}\begin{pmatrix} 2 & 1 \\ 1 & 2 \end{pmatrix},$$

which is just of inverse of the SU(3) Cartan matrix. The norms of the left and right components are,

$$w_L^{\,2} \;=\; (1/3)\ (n_1,n_2,n_3,n_4) \begin{bmatrix} 2 & 1 & 1 & 1 \\ 1 & 2 & -1 & -1 \\ 1 & -1 & 2 & 2 \\ 1 & -1 & 2 & 2 \end{bmatrix} \begin{bmatrix} n_1 \\ n_2 \\ n_3 \\ n_4 \end{bmatrix} \tag{28}$$

and the metric for $w_R^{\,2}$ is

$$(1/3) \begin{bmatrix} 2 & -2 & 1 & 1 \\ -2 & 2 & -1 & -1 \\ 1 & -1 & 2 & -1 \\ 1 & -1 & -1 & 2 \end{bmatrix} .$$

Thus, the norm of a vector w is

$$w^2 = 2n_1n_2 + 2n_3n_4 \quad , \tag{29}$$

and the lattice is even and self dual.

We now construct the spectrum of states of the closed bosonic string, reduced by two dimensions of this lattice. The calculation is a Fock space calculation of the same kind as we have often done. The mass operator is

$$\alpha' M^2 = N_L + (1/2)\, w_L{}^2 - 1,$$

with the constraint from the translation invariance of the origin being,

$$N_L + (1/2) w_L{}^2 = N_R + (1/2)\, w_R{}^2 \ .$$

The spectrum of states are:

$(1/2)\alpha' M_2$	Spin	SU(3) X SU(3)	
-1	(0000...)	(1,1)	(the usual tachyon)
-2/3	(0000...)	$(3,3) + (\overline{3},\overline{3})$	(the new tachyon)
-1/3	Nothing		
0	(2000...)	(1,1)	(graviton)
0	(1000...)	(1,8)+(8,1)	(Y. M. vectors)
0	(0100...)	(1,1)	(Antisymmetric tensor)
0	(0000...)	(1,1)+(8,8)	
	Massive representations		
1/3	(2000...)	$(3,3)+(\overline{3},\overline{3})$	
	(0100...)	$(3,3)+(\overline{3},\overline{3})$	
	(1000...)	$(6,\overline{3})+(\overline{6},3)+(\overline{3},6)+(3,\overline{6})$	
	(0000...)	$(6,6)+(\overline{6},\overline{6})+(3,3)+(\overline{3},\overline{3})$	
2/3	Nothing		
1	(4000...)	(1,1)	
	(3000...)	(1,8)+(8,1)	

(2100...)	(1,1)
(0200...)	(1,1)
(2000...)	3(1,1)+(1,8)+(8,1)+(8,8)
(1100...)	(1,8)+(8,1)
(0100...)	(1,1)+(8,8)
(1000...)	2{(1,8)+(8,1)+(8,8)}
(0000...)	2(1,1)+(1,8)+(8,1)+2(8,8)

We conclude with several comments on Lorentzian lattices. The lattice, $\Gamma_{p,q}$, which has a signature with p pluses and q minuses, can be even self dual only if $p = q + 8n$. Thus, their number appears to be greatly limited. Neverthess, if the G X G symmetry structure is not required, then it is possible to deform these lattices by a general SO(p, q) transformation. The independent parameters live in the coset SO(p, q)/SO(p)XSO(q); there are pq free parameters. An examination of $\Gamma_{1,1}$ shows how this works. The general prescription indicates that there should be a one-parameter family of solutions to the even self dual constraints. In a basis where the metric is a diagonal matrix with (1,-1), the basis vectors are

$$v_1 = (1/\sqrt{2})\, e^X\, (1,\ 1)$$

$$v_2 = (1/\sqrt{2})\, e^{-X}\, (1,\ -1)\ .$$

Note that v_1 and v_2 are null vectors, with their duals being v_2 and v_1, respectively. Thus the lattice is self dual.

The general vector on this lattice is

$$v = (1/\sqrt{2}) \left(n_1 e^{-X} + n_2\, e^X\ ,\ n_2\, e^X - n_1\, e^{-X} \right).$$

For two vectors labeled by integers n_1, n_2 and m_1, m_2, the scalar product is

$$u_1 \cdot u_2 \;=\; n_1 m_2 + n_2 m_1\ .$$

Thus, this lattice is even self dual for all values of x. However, $v_L - v_R$ is not a root of SU(2) unless $x = 0$, since,

$$v_L - v_R = \sqrt{2}\, n_1\, e^{-x} \quad .$$

This result suggests that there is a continuum of consistent string theories. (The modular invariance of the one-loop amplitudes is checked in [3].)

The parameters take on critical values when the Lorentzian lattice is composed of the weights in G X G. For other values the symmetry is reduced. It is a subject of present investigation whether or not this freedom will allow for a discussion of symmetry breaking in the S-Matrix framework.

References:

1. S. Raby and R. Slansky, Phys. Rev. Lett. Feb. (1986).

2. F. Englert and A. Neveu, Phys. Lett. **163B** (1985); A. Casher, F. Englert, H. Nicolai and A. Taormina Phys. Lett. **162B** (1985),121.

3. R. Narain, Rutherford Preprint (1986).

INTRODUCTION TO STRING FIELD THEORY

Joseph Lykken and Stuart Raby

Theoretical Division
T-8, Mail Stop B285
Los Alamos National Laboratory
Los Alamos, N.M. 87545

Abstract

We propose an action for an interacting closed bosonic string. Our formalism relies heavily on ideas discussed by Witten for the open bosonic string. We also obtain the gauge fixed quantum action for the fully interacting open bosonic string.

1 INTRODUCTION

String theories may very well provide a unified description of all the known particles and their interactions - including gravity. The 17 or more phenomenological parameters of the standard model may, in principle, be determined in this fundamental theory with no free parameters. Such are the fervent hopes of particle theorists. Yet if we are to understand this new theory we must relinquish the paradigms of the point particle, local field theory and local gauge symmetries. These concepts are presumably relevant descriptions of nature only in a low energy approximation. For example, the principle of general coordinate invariance is probably not a useful concept as applied to the string. The Virasoro-Shapiro model contains a massless spin two "graviton". However this state has no local four point interactions as is the case for the graviton in general relativity.[1] Thus the "graviton" of the Virasoro-Shapiro model is not a simple piece of a tensor under general coordinate transformations.[2] A new symmetry principle based on the string is thus needed to supersede some old but valued concepts at distance scales where stringiness becomes relevant.

Recent work towards a covariant string field theory [1-27] is partly motivated by this desire to find the symmetry principle behind the theory. We feel that a significant step in this direction has recently been taken by Witten [3] in his creation of an interacting open bosonic string field theory using some ideas of Connes [29] on non-commutative differential geometry. In this paper we explore these ideas with respect to the open bosonic string in some

[1] General relativity is only valid in the effective low energy theory defined in the zero slope limit.

[2] Perhaps a complicated field redefinition of the "graviton" in the Virasoro-Shapiro model does actually transform as $h_{\mu\nu} = g_{\mu\nu} - \eta_{\mu\nu}$ in general relativity - private communication - B. Zwiebach.

more detail. Using the intuition so obtained we propose an action for an interacting closed bosonic string field theory. In section 2 we review Witten's theory of the open bosonic string. We use a notation, however, more akin to that of Siegel and Zwiebach [2]. We present the gauge fixed BRST [30-34] invariant action for Witten's theory. This result can be found in equation (1.33). In section 3 we describe our results on the closed bosonic string. We discuss some new axioms which can lead to a gauge invariant interacting closed bosonic string action. The relevant equations are (2.20) - (2.25). We then motivate the new set of axioms. In section 4 we summarize our results and discuss the problems which remain to be explored.

2 THE OPEN BOSONIC STRING

Following Siegel [1] we define a real string functional

$$\Phi[x_\mu(\sigma), c(\sigma), \tilde{c}(\sigma)], \tag{1.1}$$

and a BRST generator Q, satisfying $Q^2 = 0$ [35,36], represented as a differential operator acting on Φ – –

$$Q = \sqrt{\pi\alpha'}\int_{-\pi}^{\pi} d\sigma \hat{c}(\sigma)[\frac{1}{2}\hat{P}^2(\sigma) - \frac{i}{\alpha'}\hat{c}'(\sigma)\Pi_{\hat{c}}(\sigma)) \tag{1.2}$$

where

$$\begin{aligned} \hat{P}_\mu(\sigma) &= 1/\sqrt{2}[i\frac{\delta}{\delta x^\mu(\sigma)} - \frac{1}{\alpha'}x'_\mu(\sigma)], \\ \hat{c}(\sigma) &= \alpha'/\sqrt{2}[\frac{\delta}{\delta c(\sigma)} + \frac{1}{\alpha'}\tilde{c}(\sigma)], \\ \Pi_{\hat{c}}(\sigma) &= 1/\sqrt{2}[\frac{\delta}{\delta \tilde{c}(\sigma)} + \frac{1}{\alpha'}c(\sigma)]. \end{aligned} \tag{1.3}$$

The commuting (anti-commuting) coordinates $x_\mu(\sigma)(c(\sigma), \tilde{c}(\sigma))$ with $\mu = 1, \cdots, 26$ and $0 \le \sigma \le \pi$ satisfy the boundary conditions $x'_\mu(\sigma) \equiv \frac{dx_\mu(\sigma)}{d\sigma} = 0, c'(\sigma) = 0, \tilde{c}(\sigma) = 0$ at $\sigma = 0, \pi$. (The coordinates for the region $-\pi \le \sigma \le 0$ are then obtained by analytic continuation. α' is the string tension.) The string functional Φ can be expanded in terms of an infinite number of local component fields. This expansion is particularly useful when making contact with local field theory. We define the differential operators $a_{n\mu}, \beta_n, \tilde{\beta}_n$ for $-\infty \le n \le \infty$ as the normal modes of the operators given in (eq. 1.3)-

$$\begin{aligned} \hat{P}_\mu(\sigma) &= 1/\sqrt{2\pi\alpha'}\sum a_{n\mu}e^{-in\sigma}, \\ \hat{c}(\sigma) &= \sqrt{\alpha'/2\pi}\sum \beta_n e^{-in\sigma}, \\ \Pi_{\hat{c}}(\sigma) &= 1/\sqrt{2\pi\alpha'}\sum \tilde{\beta}_n e^{-in\sigma} \end{aligned} \tag{1.4}$$

They satisfy the canonical commutation and anti-commutation relations

$$\begin{aligned} [a_{n\mu}, a_{m\nu}] &= n\delta_{n+m,0}\,\eta_{\mu\nu} \\ \{\beta_n, \tilde{\beta}_m\} &= \delta_{n+m,0} \end{aligned} \tag{1.5}$$

We also define a ground state functional Φ_0 satisfying

$$a_{n\mu}\Phi_0 = \beta_n\Phi_0 = \tilde{\beta}_m\Phi_0 = 0 \text{ for } n \ge 0, m > 0. \tag{1.6}$$

The component expression is then given by $\Phi = \{i\tilde{\beta}_1^\dagger B(x) + \cdots + c(\phi(x) + a_{1\mu}^\dagger A^\mu(x) + \beta_1^\dagger \bar{\eta}(x) + \tilde{\beta}_1^\dagger \eta(x) + \cdots)\}\Phi_0$ where

$$c \equiv \sqrt{\alpha'/\pi}\tilde{\beta}_0 \text{ and } \partial_c \equiv \sqrt{\pi/\alpha'}\beta_0 \tag{1.7}$$

Φ is a commuting field. Thus the fields ϕ , A^μ ,B are also commuting and correspond to the tachyon, abelian gauge field and Lautrup-Nakanishi auxiliary field. η and $\bar{\eta}$ are the anti-commuting Fadeev-Popov ghost and anti-ghost fields connected with A^μ and B. A global BRST transformation of Φ is given by

$$\delta\Phi = i\epsilon Q\Phi \tag{1.8}$$

which leaves the free action

$$S_0 = \int \Phi^T[Q\,,\rho]\Phi \tag{1.9}$$

invariant. ($\Phi^T \equiv \Phi[x_\mu(\pi-\sigma), c(\pi-\sigma), -\tilde{c}(\pi-\sigma)]$) ρ is the gauge fixing operator and the Feynman-Siegel gauge is obtained by the choice

$$\rho = [c, \partial_c]. \tag{1.10}$$

A particularly simple subspace to consider , for illustrative purposes, is that including the massless sector alone i.e., A^μ , B , η and $\bar{\eta}$. In this subspace the BRST generator Q assumes the form

$$Q_0 = [\beta_1 a_1^\dagger.i\partial + \beta_1^\dagger a_1.i\partial - \partial_c\Box - \beta_1^\dagger\beta_1 c] \tag{1.11}$$

and satisfies $Q_0^2 = 0$ (we have taken $\alpha' = 2\pi$). It is then trivial to check that the BRST transformation (eq. 1.8) for the massless fields is given by

$$\begin{aligned} \delta A^\mu &= -\epsilon\partial^\mu\eta \\ \delta\eta &= 0 \\ \delta B &= \epsilon\Box\eta \\ \delta\bar{\eta} &= \epsilon(\partial.A + B). \end{aligned} \tag{1.12}$$

Let us now consider the local gauge invariant theory of Witten. We must first define the physical subspace of Φ. Following Witten we consider the ghost number operator given by

$$N_g = \frac{1}{2}[\partial_c\,, c] + \sum_{n>0}(\beta_n^\dagger\tilde{\beta}_n - \tilde{\beta}_n^\dagger\beta_n). \tag{1.13}$$

Then $N_g\Phi_0 = 1/2\,\Phi_0$ and the physical sector is defined to have ghost number $-\frac{1}{2}$. (We shall denote by capital Latin letters string functionals satisfying the constraint $N_g A = -\frac{1}{2}A$.) The component expansion is given by

$$A = \{i\tilde{\beta}_1^\dagger B(x) + \cdots + c(\phi(x) + a_{1\mu}^\dagger A^\mu(x) + \cdots)\}\Phi_0. \tag{1.14}$$

The free field equations are then given by

$$QA = 0 \tag{1.15}$$

which is invariant under the local gauge transformation

$$\delta A = Q\epsilon. \tag{1.16}$$

Note that $[N_g\,, Q] = Q$ and thus ϵ must have ghost number $-\frac{3}{2}$. In the massless subsector we find eqs. 1.15 and 1.16 imply

$$\begin{aligned} \Box A_\mu + \partial_\mu B &= 0\,, \partial.A + B = 0 \\ \delta A_\mu &= \partial_\mu\lambda\,, \delta B = -\Box\lambda \end{aligned} \tag{1.17}$$

where $\epsilon = \{ic\tilde{\beta}_1^\dagger\lambda(x) + \cdots\}\Phi_0$. Eqs. 1.17 are equivalent to the free Maxwell equations

$\partial^{\mu} F_{\mu\nu} = 0$. The free action S_{W0} is given by

$$S_{W0} = \int A^T Q A. \qquad (1.18)$$

In order to generalize this theory to include interactions Witten relies on the axioms of the non-commutative differential geometry of Connes. In the rest of this section we shall

(1) review these axioms , and

(2) discuss an explicit representation of them.

2.1 Abstract Axioms

We want the string functionals A to satisfy properties analogous to those of forms in differential geometry and for the BRST generator Q to act as a derivation. Q already satisfies the first axiom

(i) $Q^2 = 0$.

We define a product rule denoted by $*$ and an integral $\int$ which satisfy:

(ii) associativity of $*$ product-

$(A * B) * C = A * (B * C)$.

Although the product is non-commutative $(A * B \neq B * A)$, we demand

(iii) $\int A * B = (-1)^{AB} \int B * A$

where $(-1)^{AB}$ = (-1, if A and B are Grassman odd; +1, otherwise). The additional properties of a derivation are:

(iv) $\int Q\chi = 0$ for any string functional χ and, Q must satisfy

(v) the Leibnitz rule-

$Q(A * B) = QA * B + (-1)^A A * QB$ where $(-1)^A$ = (-1, if A is Grassman odd; +1, otherwise).

Given these properties we can then define the gauge field strength

$$F = QA + A * A \qquad (1.19)$$

which under the local gauge transformation

$$\delta A = Q\epsilon + [A , \epsilon] \qquad (1.20)$$

(where$[A, \epsilon] \equiv A * \epsilon - \epsilon * A$) transforms by

$$\delta F = [F , \epsilon]. \qquad (1.21)$$

[Note that the group structure implicit in this abstract algebra is encoded in the $*$product. For example it is easy to see that the transformation of A defined in eq. 1.20 satisfies the group property

$$[\delta_2 , \delta_1] A = \delta_3 A \qquad (1.22)$$

where the infinitesimal gauge functional parameters ϵ_i, i=1,2,3 are related by $\epsilon_3 = [\epsilon_2, \epsilon_1]$.]

Finally the action for the interacting theory is given by

$$S_W = \int (A * QA + \frac{2}{3} A * A * A) \tag{1.23}$$

which is invariant under (1.20). The equations of motion are given by

$$\delta S/\delta A = F = 0. \tag{1.24}$$

The ghost numbers of $\int$ and $*$ were determined by Witten to be $-\frac{3}{2}$ and $+\frac{3}{2}$, respectively. Moreover $\int \chi = 0$ unless χ has ghost number $+\frac{3}{2}$. Hence a non-trivial action has net ghost number zero.

2.2 Representation of the Axioms

Witten has shown that an associative $*$ product requires treating the point $\sigma = \pi/2$ special. Thus A$*$B is defined by a process of "sewing" the right half of string A with the left half of string B as depicted in fig.1.

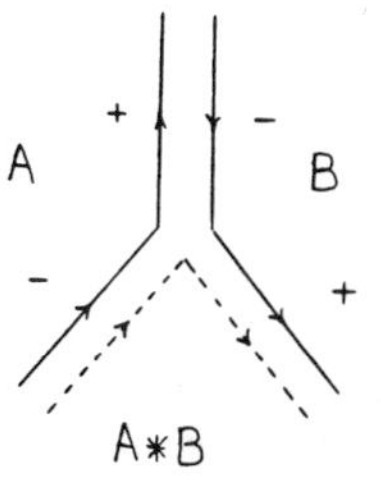

Fig. 1

This statement will be made more precise shortly. If A$*$B is given by fig.1, then integral $\int$ might be expected to "sew" the remaining halves of the string to each other, as in fig.2.

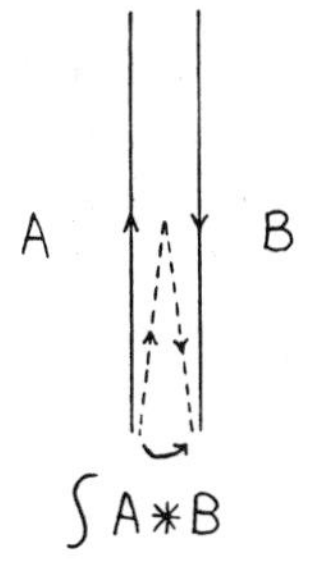

Fig. 2

In this way we obtain the "**sewing condition**"

$$\int A^T B(eq.1.18) = \int A * B \tag{1.25}$$

where $A^T \equiv A[x_\mu(\pi - \sigma), c(\pi - \sigma), -\bar{c}(\pi - \sigma)]$.

2.3 The Integral - $\int$

We shall now obtain an explicit representation of $\int$. The two main properties it must satisfy are Axiom (iv) and ghost number $-\frac{3}{2}$. The "naive" integral is represented by

$$\int = \int D\delta \tag{1.26}$$

where $D \equiv \Pi_{\sigma=0}^{\pi} dx_\mu(\sigma) dc(\sigma) d\tilde{c}(\sigma)$, $\delta \equiv \Pi_{\sigma=0}^{\pi/2} \delta(x^o)\delta(c^o)\delta(\tilde{c}^e)$ and

$$\begin{aligned} x_\mu^o(\sigma) &\equiv x_\mu(\sigma) - x_\mu(\pi - \sigma) \\ c^o(\sigma) &\equiv c(\sigma) - c(\pi - \sigma) \\ \tilde{c}^e(\sigma) &\equiv \tilde{c}(\sigma) + \tilde{c}(\pi - \sigma). \end{aligned} \tag{1.27}$$

The ghost number of the factors are D [1/2], δ [0]. Thus the "naive" integral has the wrong ghost number. Moreover we find that $\int D\delta Q\chi \neq 0$ in general, due to complications associated with the special point $\sigma = \pi/2$. Thus an insertion is needed.

The unique integral with ghost number $-\frac{3}{2}$ and satisfying Axiom (iv) is given by

$$\int = \int D\delta I \tag{1.28}$$

where $I \equiv c(\pi/2)\frac{\delta}{\delta\tilde{c}(\pi/2)}$. We then find $\int Q\chi = 0$ only in 26 dimensions.

2.4 The $*$ product

The most difficult constraint the $*$product must satisfy is the Leibnitz rule. Recall that the BRST charge Q (eq.1.2) contains products of up to three derivatives. A "naive" definition of the $*$ product is given by the expression (see fig.1)

$$\begin{aligned} A * B &= \int \Pi_{\sigma=\frac{\pi}{2}}^{\pi} dx_\mu^1(\sigma) dc^1(\sigma) d\tilde{c}^1(\sigma) \Pi_{\sigma=0}^{\pi/2} dx_\mu^2(\sigma) dc^2(\sigma) d\tilde{c}^2(\sigma) \\ &\quad \times \Pi_{\sigma=0}^{\pi/2} \delta(x^2(\sigma) - x^1(\pi-\sigma))\delta(c^2(\sigma) - c^1(\pi-\sigma))\delta(\tilde{c}^2(\sigma) + \tilde{c}^1(\pi-\sigma)) \\ &\quad A[1]B[2] \\ &\equiv \int \mu^* AB \end{aligned} \tag{1.29}$$

(where we use the shorthand notation $A[1] = A[x_\mu^1(\sigma), c^1(\sigma), \tilde{c}^1(\sigma)]$). Since Q is defined by an integral over a local function of σ, it is possible to define a new charge Q' with the point $\sigma = \pi/2$ removed. It is trivial to show that Q' satisfies a Leibnitz rule on the "naive" $*$ product. Consider $Q'(A * B) = \int \mu^*[(Q'_- A)B + (-1)^A A(Q'_+ B)]$ where $Q'_\pm$ are defined by integrals over the $\pm$ sectors of strings A or B (see fig. 1). However, using the following identity, which is a consequence of integration by parts, $\int \mu^*[(Q'_+ A)B + (-1)^A A(Q'_- B)] \equiv 0$ and $Q'A \equiv (Q'_+ + Q'_-)A$ we find $Q'(A * B) = Q'A * B + (-1)^A A * Q'B$. The difficulty in satisfying the Leibnitz rule is thus all contained in the treatment at the special point $\sigma = \pi/2$.

We have not yet solved this difficult problem. Nevertheless we may state some general properties the $*$ product must satisfy in a notation which will be useful later in our discussion of the closed string. In general we may define a $*$ product by the expression

$$A * B[3] = \int D_1 D_2 K[312] A[1] B[2] \tag{1.30}$$

where $D_i \equiv \Pi_{\sigma=0}^{\pi} dx_\mu^i(\sigma) dc^i(\sigma) d\tilde{c}^i(\sigma)$ and the Kernel K must satisfy

$$\text{"sewing condition"}(eq.1.25) - \int D_3 \delta_3 I_3 K[312] \equiv \delta[12]; \tag{1.31a}$$

$$\text{Leibnitz rule} - \sum_{i=1}^{3} Q_i K[312] = 0, \text{and} \tag{1.31b}$$

$$\text{associativity} - \int D_3 K[534] K[312] = -\int D_3 K[513] K[324]. \tag{1.31c}$$

Property (a) fixes the ghost number of K to be $+\frac{1}{2}$ for the open string and hence the ghost number of the $*$ product is $+\frac{3}{2}$. It can also be used to show that K is Grassman even, using D_3 [odd] , δ_3 [even] , I_3 [even] and $\delta[12]$ [odd].

Once the $*$ product (or the Kernel K) is determined we may then obtain the interacting field equations in component form. Work in this direction is still in progress. Throughout the remainder of this section we shall assume that there exists a K satisfying (a-c).

The $*$ product is also significant on a deeper level. It defines a sort of matrix multiplication where the string functional A (a connection on some fibre bundle) can be thought of as an infinite dimensional matrix. Gauge transformations δA (1.20) , as usual, describe the motion of A along the fibre. However the base manifold of this fibre bundle is trivial; it's just a point.

We end this section with a discussion of the gauge fixed quantum action for the fully interacting theory. Consider first the free field case. Witten's gauge invariant action $S_{W0} = \int A{*}QA$ is invariant under $\delta A = Q\epsilon$. A has ghost number $-\frac{1}{2}$. The quantum string functional (where by "quantum" we mean it includes the BRST ghost and anti-ghost fields which are necessary for a quantum description of the theory) $\Phi \equiv A + \Xi$ where the field Ξ includes all ghost numbers, excluding $-\frac{1}{2}$. Siegel's gauge fixed action may then be rewritten using the notation of $\int$ and $*$ defined previously. We have

$$\begin{aligned} S_0 &= \int \Phi * [Q,\rho]\Phi \\ &= \int A * QA + \int A * ([Q,\rho] - Q)A + \int \Xi * [Q,\rho]\Xi \end{aligned} \tag{1.32}$$

where the three terms may be identified as the gauge invariant action, the gauge fixing term and the Fadeev-Popov ghost action, respectively. S_0 is invariant under global BRST transformations $\delta\phi = i\epsilon Q\Phi$. Note the physical sector transforms according to $\delta A = Q(-i\epsilon\Xi|_{-3/2})$.

2.5 BRST Invariant Action

Let us now generalize Siegel's free gauge fixed quantum action to the fully interacting theory. We contend that

$$S_{qu} = \int (\Phi * [Q,\rho]\Phi + 2\rho\Phi * \Phi * \Phi) \tag{1.33}$$

is invariant under the non-linear global BRST transformation

$$\delta\Phi = i\epsilon(Q\Phi + \Phi * \Phi) \tag{1.34}$$

satisfying $\delta^2 = 0$. The physical sector obeys

$$\delta A = Q\xi + [A,\xi] + i\epsilon\Xi^*_{n_1}\Xi_{n_2}|_{n_1+n_2=-2} \tag{1.35}$$

where $\xi = -i\epsilon\Xi|_{-\frac{3}{2}}$. Except for the last term, this is just a generalization of global BRST transformations for non-abelian gauge theories.

3 THE CLOSED BOSONIC STRING

Following Siegel (once more) we define a real closed string functional

$$\Phi[x_\mu(\sigma), c(\sigma), \tilde{c}(\sigma)] \tag{2.1}$$

and a BRST generator Q, satisfying $Q^2 = 0$. Q is given by

$$\begin{aligned} Q &= \sqrt{2\pi\alpha'}\int_{-\pi}^{\pi} d\sigma\{\hat{c}(\sigma)[\frac{1}{2}\hat{P}^2(\sigma) - \frac{i}{\alpha'}\hat{c}'(\sigma)\Pi_{\hat{c}}(\sigma)] \\ &+ \bar{\hat{c}}(\sigma)[\frac{1}{2}\bar{\hat{P}}^2(\sigma) + \frac{i}{\alpha'}\bar{\hat{c}}'(\sigma)\bar{\Pi}_{\hat{c}}(\sigma)]\} \end{aligned} \tag{2.2}$$

where

$$\begin{aligned} \hat{P}_\mu(\sigma) &= 1/\sqrt{2}[i\frac{\delta}{\delta x_\mu(\sigma)} - \frac{1}{\alpha'}x'_\mu(\sigma)], \\ \hat{c}(\sigma) &= \alpha'/\sqrt{2}[\frac{\delta}{\delta c(\sigma)} + \frac{1}{\alpha'}\tilde{c}(\sigma)], \\ \Pi_{\hat{c}}(\sigma) &= 1/\sqrt{2}[\frac{\delta}{\delta\tilde{c}(\sigma)} + \frac{1}{\alpha'}c(\sigma)]. \\ \bar{\hat{P}}_\mu(\sigma) &= 1/\sqrt{2}[i\frac{\delta}{\delta x_\mu(\sigma)} + \frac{1}{\alpha'}x'_\mu(\sigma)], \\ \bar{\hat{c}}(\sigma) &= \alpha'/\sqrt{2}[\frac{\delta}{\delta c(\sigma)} - \frac{1}{\alpha'}\tilde{c}(\sigma)], \\ \bar{\Pi}_{\hat{c}}(\sigma) &= 1/\sqrt{2}[-\frac{\delta}{\delta\tilde{c}(\sigma)} + \frac{1}{\alpha'}c(\sigma)]. \end{aligned} \tag{2.3}$$

The commuting (anti-commuting) coordinates $x_\mu(\sigma)(c(\sigma), \tilde{c}(\sigma))$ with $\mu = 1, \cdots, 26$ and $-\pi \leq \sigma \leq \pi$ satisfy the periodic boundary conditions $x_\mu(\sigma) = x_\mu(\sigma + 2\pi), c(\sigma) = c(\sigma + 2\pi), \tilde{c}(\sigma) = \tilde{c}(\sigma + 2\pi)$. Note that $\tilde{c}(\sigma)$ now has a zero mode, unlike the case of the open string.

As in the case of the open string, we may expand Φ in terms of an infinite number of local component fields. We define the differential operators $a_{n\mu}, \beta_n, \tilde{\beta}_n, \bar{a}_{n\mu}, \bar{\beta}_n, \bar{\tilde{\beta}}_n$ for $-\infty \leq n \leq \infty$ by

$$\begin{aligned} \hat{P}_\mu(\sigma) &= 1/\sqrt{2\pi\alpha'}\sum a_{n\mu}e^{-in\sigma}, \\ \hat{c}(\sigma) &= \sqrt{\alpha'/2\pi}\sum \beta_n e^{-in\sigma}, \\ \Pi_{\hat{c}}(\sigma) &= 1/\sqrt{2\pi\alpha'}\sum \tilde{\beta}_n e^{-in\sigma}, \\ \bar{\hat{P}}_\mu(\sigma) &= 1/\sqrt{2\pi\alpha'}\sum \bar{a}_{n\mu}e^{in\sigma}, \\ \bar{\hat{c}}(\sigma) &= \sqrt{\alpha'/2\pi}\sum \bar{\beta}_n e^{in\sigma}, \\ \bar{\Pi}_{\hat{c}}(\sigma) &= 1/\sqrt{2\pi\alpha'}\sum \bar{\tilde{\beta}}_n e^{in\sigma} \end{aligned} \tag{2.4}$$

satisfying the canonical commutation and anti-commutation relations

$$\begin{aligned} [a_{n\mu}, a_{m\nu}] &= n\delta_{n+m,0}\,\eta_{\mu\nu}, \\ \{\beta_n, \tilde{\beta}_m\} &= \delta_{n+m,0}, \\ [\bar{a}_{n\mu}, \bar{a}_{m\nu}] &= n\delta_{n+m,0}\,\eta_{\mu\nu}, \\ \{\bar{\beta}_n, \bar{\tilde{\beta}}_m\} &= \delta_{n+m,0} \end{aligned} \tag{2.5}$$

with all others zero. We also define a ground state functional Φ_0 satisfying

$$a_{n\mu}\Phi_0 = \beta_m\Phi_0 = \tilde{\beta}_m\Phi_0 = \bar{a}_{n\mu}\Phi_0 = \bar{\beta}_m\Phi_0 = \bar{\tilde{\beta}}_m\Phi_0 = 0 \text{ for } n \geq 0, m > 0. \qquad (2.6)$$

We also have

$$\partial_c\Phi_0 = \partial_{\tilde{c}}\Phi_0 = 0 \qquad (2.7)$$

where $\partial_c \equiv \sqrt{\pi/\alpha'}(\beta_0 + \bar{\beta}_0)$ and $\partial_{\tilde{c}} \equiv \sqrt{\pi/\alpha'}(\tilde{\beta}_0 - \bar{\tilde{\beta}}_0)$.

The component expansion of Φ includes a tachyon, massless graviton, Kalb-Ramond anti-symmetric tensor, dilaton and an infinite number of massive fields. It also includes the necessary Fadeev-Popov ghosts, auxiliary fields and many more unphysical fields.

Global BRST transformations are again given by

$$\delta\Phi = i\epsilon Q\Phi. \qquad (2.8)$$

We now want to discuss a gauge invariant description for the closed bosonic string. The physical sector of Φ may be defined a la Witten by its ghost number content. We define the ghost number operator

$$N_g = \frac{1}{2}[\partial_c, c] - \frac{1}{2}[\partial_{\tilde{c}}, \tilde{c}] + \sum_{n>0}(\beta_n^\dagger\tilde{\beta}_n - \tilde{\beta}_n^\dagger\beta_n + \bar{\beta}_n^\dagger\bar{\tilde{\beta}}_n - \bar{\tilde{\beta}}_n^\dagger\bar{\beta}_n) \text{ with } N_g\Phi_0 = 0. \quad (2.9)$$

Then the physical sector is defined to have ghost number -1. (We shall use capital Latin letters for physical string functionals which satisfy $N_g A = -A$.) We shall impose two additional constraints on Φin order to define the physical sector. The first is the standard requirement that the string functional is invariant under constant reparametrizations of σ i.e. the "origin" of the string has no physical significance, or

$$A[x_\mu(\sigma + f), c(\sigma + f), \tilde{c}(\sigma + f)] = A[x_\mu(\sigma), c(\sigma), \tilde{c}(\sigma)] \qquad (2.10)$$

for constant shift f. Finally we constrain the dependence of A on the ghost zero mode $\tilde{c}$. We demand that A be expressible in the form $A = \partial_{\tilde{c}}\chi$ which implies $\partial_{\tilde{c}}A \equiv 0$. To summarize, the physical sector of Φ is defined to satisfy

$$N_g A = -A, \; \Delta N A = 0, \; A = \partial_{\tilde{c}}\chi \qquad (2.11)$$

where $\Delta N \equiv N - \bar{N}$ (the number of right minus left movers of the string) is the generator of constant σ reparametrizations.

The free field equations of motion are given by

$$QA = 0 \qquad (2.12)$$

with the corresponding gauge invariance

$$\delta A = Q\epsilon. \qquad (2.13)$$

The component expansion of A is given by

$$\begin{aligned} A \;=\; & \{i\tilde{\beta}_1^\dagger\bar{a}_{1\mu}^\dagger\bar{B}^\mu(x) + i\bar{\tilde{\beta}}_1^\dagger a_{1\mu}^\dagger B^\mu(x) + \cdots + c(\phi(x) + \\ & \bar{a}_{1\mu}^\dagger a_{1\nu}^\dagger h^{\mu\nu}(x) + \bar{\beta}_1^\dagger\tilde{\beta}_1^\dagger\bar{\chi}(x) + \bar{\tilde{\beta}}_1^\dagger\beta_1^\dagger\chi(x) + \cdots)\}\Phi_0. \end{aligned} \qquad (2.14)$$

In the massless subspace the BRST generator takes the simple form

$$Q_0 = [\beta_1 a_1^\dagger.i\partial + \beta_1^\dagger a_1.i\partial + \bar{\beta}_1\bar{a}_1^\dagger.i\partial + \bar{\beta}_1^\dagger\bar{a}_1.i\partial - \partial_c\Box - (\beta_1^\dagger\beta_1 + \bar{\beta}_1^\dagger\bar{\beta}_1)c] \qquad (2.15)$$

satisfying $Q_0^2 = 0$ (with $\alpha' = 4\pi$). It is then trivial to check that in the massless subsector eqs. 2.12 and 2.13 give the equations of motion

$$\begin{aligned}
\Box h^{\mu\nu} + \partial^\nu \bar{B}^\mu + \partial^\mu B^\nu &= 0\,, \\
\Box\chi - \partial.B = 0\,,\ \Box\bar{\chi} + \partial.\bar{B} &= 0 \\
\partial_\nu h^{\mu\nu} + \bar{B}^\mu + \partial^\mu\chi &= 0\,, \\
\partial_\mu h^{\mu\nu} + B^\nu - \partial^\nu\bar{\chi} &= 0;
\end{aligned} \tag{2.16}$$

and local gauge transformations

$$\begin{aligned}
\delta h^{\mu\nu} &= \partial^\nu\bar{\xi}^\mu + \partial^\mu\xi^\nu \\
\delta\bar{B}^\mu &= -\Box\bar{\xi}^\mu + \partial^\mu\eta,\ \delta B^\nu = -\Box\xi^\nu - \partial^\nu\eta \\
\delta\bar{\chi} &= \partial.\bar{\xi} - \eta,\ \delta\chi = -\partial.\xi - \eta
\end{aligned} \tag{2.17}$$

where $\epsilon = \{\bar{\tilde{\beta}}_1^\dagger\tilde{\beta}_1^\dagger\eta(x) + ic(\tilde{\beta}_1^\dagger\bar{a}_{1\mu}^\dagger\bar{\xi}^\mu(x) + \bar{\tilde{\beta}}_1^\dagger a_{1\mu}^\dagger\xi^\mu(x)) + \cdots\}\Phi_0$.

These are just the equations of motion for linearized gravity, the Kalb-Ramond gauge field and a massless dilaton.

An action may be constructed for the free theory. We find[3]

$$S_0 = \int A^T \frac{1}{2}[\tilde{c}, Q]A \tag{2.18}$$

where $A^T \equiv A[x_\mu(-\sigma), c(-\sigma), -\tilde{c}(-\sigma)]$. Note the insertion of the ghost zero mode coordinate $\tilde{c}$. This is necessary; otherwise the integral would vanish upon $\tilde{c}$ integration. The equations of motion derived from this action are

$$\partial_{\tilde{c}}[\tilde{c}, Q]A = 0 \tag{2.19}$$

which (using the constraints eq. (2.11)) is equivalent to eq. (2.12). [Note that Q may generally be written in the form $Q = Q' + \sqrt{\frac{\pi}{\alpha'}}\Delta N\ \tilde{c} + \sqrt{\frac{\alpha'}{\pi}}\Delta T\ \partial_{\tilde{c}}$ where we have made the $\tilde{c}$ dependence explicit. It is then trivial to prove the useful identity $QA \equiv \frac{1}{2}\partial_{\tilde{c}}[\tilde{c}, Q]A$.]

Let us now discuss our proposal for the fully interacting closed bosonic string theory. We want to write an action which is similar to that for the open bosonic string (eq. 1.23). In order to accomplish this we must change some of the axioms of section 2. We shall assume that we can define an integral $\int$ and $*$ product which satisfy the following axioms. Although we have not yet completely defined a representation for the new axioms, we shall try to motivate each one by comparison with our knowledge of the open string. It is not inconceivable to us that the axioms we discuss may, in fact, already be represented by the path integral expressions defined by Witten.

3.1 Abstract Axioms

(i) $Q^2 = 0$

We define a product rule which is Grassman odd and satisfies $\Delta N(A * B) = 0$ if $\Delta NA = \Delta NB = 0$, but $\partial_{\tilde{c}}A * B \neq 0$, even if $\partial_{\tilde{c}}A = \partial_{\tilde{c}}B = 0$. The "associativity" axioms of the $*$ product are

(ii) **"associativity"-**

$$\begin{aligned}
(A * B) * C &= -(-1)^A A * (B * C) \\
\partial_{\tilde{c}}(\partial_{\tilde{c}}(A * B) * C) &= \partial_{\tilde{c}}(A * \partial_{\tilde{c}}(B * C)).
\end{aligned}$$

[3] A similar result for the free closed bosonic string has been obtained by Banks et al. [13].

The integral is defined to satisfy

(iii) $\int A * B = -(-1)^{AB} \int B * A$.

Q satisfies the additional axioms

(iv) $\int Q\chi = 0$ for any string functional χ, and

(v) the **"Leibnitz rule"-**

$-Q(A * B) = QA * B + (-1)^A A * QB$.

Note that the axioms (i) and (iv) are identical to those of the open string. (ii), (iii) and (v) have been altered. We shall also assume that the integral $\int$ and $*$ product can be defined to have ghost number -3 and +3, respectively.

Before we attempt to justify our apparently bizarre set of axioms, let us first discuss some results which follow from their use. We are able to define a field strength

$$F = \partial_{\tilde{c}}(\frac{1}{2}[\tilde{c}, Q]A + A * A) \equiv QA + \partial_{\tilde{c}}(A * A) \quad (2.20)$$

which under the local gauge transformation

$$\delta A = Q\epsilon + \partial_{\tilde{c}}[A, \epsilon] \quad (2.21)$$

(where $[A, \epsilon] \equiv A * \epsilon - \epsilon * A$) transforms by

$$\delta F = \partial_{\tilde{c}}[F, \epsilon]. \quad (2.22)$$

The fully interacting theory is then defined by the action

$$S = \int (A * \frac{1}{2}[\tilde{c}, Q]A + \frac{2}{3} A * A * A). \quad (2.23)$$

[Note that as a result of the axioms $\int (A * A) * A = \int A * (A * A)$.] It is invariant under the transformation (2.21). A proof of this statement requires using the following identities which follow from integration by parts and the "sewing condition":

$$\text{"sewing condition"} - \int A * B = \int A^T B \quad (2.24)$$

(where A^T is defined after eq.(2.18))

identities -

$$\begin{aligned} \int (\partial_{\tilde{c}}A) * B &= (-1)^A \int A * \partial_{\tilde{c}}B \\ \int A * [\tilde{c}, Q]B &= \int ([\tilde{c}, Q]A) * B. \end{aligned}$$

The equations of motion are given by

$$\frac{\delta S}{\delta \chi} = F = 0 \quad (2.25)$$

where $A \equiv \partial_{\tilde{c}}\chi$. For the free theory these equations reduce to those of eq.(2.12) which has been shown to include the equations of linearized gravity.

4 REPRESENTATION OF THE AXIOMS

4.1 The Integral - $\int$

Our definition of the integral for the closed string is a simple generalization of our results for the open string. The "naive" integral is defined by folding the closed loop in half (see fig. 3) and "sewing" the two halves together.

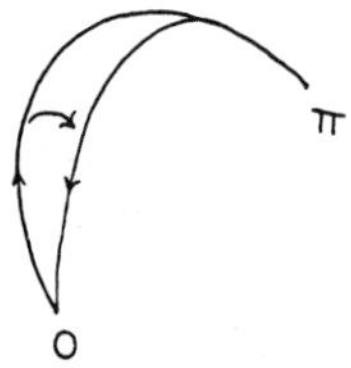

Fig. 3

There are now two singular points at $\sigma = 0$ and π, which requires an insertion in order to satisfy axiom (iv). The correct measure is given by

$$\int \equiv \int D\delta I \tag{2.26}$$

where

$$\begin{aligned} D &\equiv \Pi_{\sigma=-\pi}^{\pi} dx_\mu(\sigma)\, dc(\sigma)\, d\tilde{c}(\sigma)\,, \delta \equiv \Pi_{\sigma=0}^{\pi} \delta(x^o)\,\delta(c^o)\,\delta(\tilde{c}^e)\,, \\ x_\mu^o(\sigma) &\equiv x_\mu(\sigma) - x_\mu(-\sigma)\,, \\ c^o(\sigma) &\equiv c(\sigma) - c(-\sigma)\,, \\ \tilde{c}^e(\sigma) &\equiv \tilde{c}(\sigma) + \tilde{c}(-\sigma)\,, \end{aligned}$$

and $I \equiv c(0)\,\frac{\delta}{\delta\tilde{c}(0)} \times c(\pi)\,\frac{\delta}{\delta\tilde{c}(\pi)}$. The ghost number of the factors are $D[0]$, $\delta[1]$ and $I[-4]$. Thus the integral has ghost number -3. It satisfies $\int Q\chi = 0$ for all χ (when the dimension of space-time is 26).

4.2 The $*$ product

We shall define a $*$ product by the expression

$$A * B[3] = \int D_1 D_2 K[312] A[1] B[2] \tag{2.27}$$

where $D_i \equiv \Pi_{\sigma=-\pi}^{\pi} dx_\mu^i(\sigma)\, dc^i(\sigma)\, d\tilde{c}^i(\sigma)$. The "naive" kernel K is a product of delta functions which "sew" together half of string A to half of string B as shown in fig. 4.

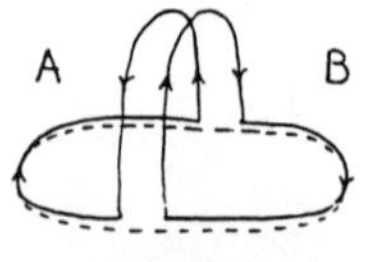

Fig. 4

We assume that the integral and $*$ product satisfy the "sewing condition" (eq. 2.24) which implies that K satisfies

$$\int D_3 \delta_3 I_3 K[312] \equiv \delta[12]. \qquad (2.28)$$

The "sewing condition" determines two important properties of K and thus of the $*$ product. It fixes the ghost number of K to be +3, since $\delta[12]$ has zero ghost number. Thus the $*$ product has ghost number +3, as desired. It also fixes the kernel K to be Grassman [odd], using D_3 [even], δ_3 [odd], I_3[even] and $\delta[12]$ [even]. Most of the changes in the axioms can be traced back to the fact that for the closed string the kernel is Grassman [odd] , where as for the open string K is Grassman [even]. To illustrate why this change occurs consider one of the factors, for example D. For the open string $D \equiv \Pi_{\sigma=0}^{\pi} dx_\mu(\sigma) dc(\sigma) d\tilde{c}(\sigma)$ which in component form is given by $D \equiv \Pi_{n=0}^{\infty} dx_{n\mu} \Pi_{n=0}^{\infty} dc_n \Pi_{n=1}^{\infty} d\tilde{c}_n$. The product $\Pi_{n=1}^{\infty} dc_n \times d\tilde{c}_n$ is Grassman even; thus the extra zero mode for $c(\sigma)$ makes D Grassman odd. For the closed string $D \equiv \Pi_{\sigma=-\pi}^{\pi} dx_\mu(\sigma) dc(\sigma) d\tilde{c}(\sigma)$ which in component form is given by $D \equiv \Pi_{n=-\infty}^{+\infty} dx_{n\mu} \Pi_{n=-\infty}^{+\infty} dc_n \Pi_{n=-\infty}^{+\infty} d\tilde{c}_n$. Since now there are an equal number of modes for c and $\tilde{c}$, D is Grassman even. Similar considerations have been used to obtain the Grassmanality of all the other factors. Since K is Grassman odd, one immediately concludes that A$*$A is also Grassman odd.

Let us now reconsider the axioms.

(ii) **"associativity"** -

The expression (A$*$B)$*$C = $-(-1)^A A$$*$(B$*$C) is equivalent to the following familiar relation for K : $\int D_3 K[534] K[312] = -\int D_3 K[513] K[324]$ (see eq.1.31c for the open string). Since K is Grassman odd it is not possible for the $*$ product (eq.2.27) to be associative for both even and odd Grassman valued fields.

The expression $\partial_{\tilde{c}}(\partial_{\tilde{c}}(A*B)*C) = \partial_{\tilde{c}}(A*\partial_{\tilde{c}}(B*C))$ is equivalent to the relation for K: $\int D_3' K_1[534] K_1[312] = -\int D_3' K_1[513] K_1[324]$ where $K_1[312] \equiv -\partial_{\tilde{c}_3} \partial_{\tilde{c}_1} \partial_{\tilde{c}_2} K[312]$ and $D \equiv D' \partial_{\tilde{c}}$. Note since K_1 is Grassman even, the above associativity rule does not distinguish between even and odd Grassman valued fields.

(iii) $\int A * B = -(-1)^{AB} \int B * A$. This axiom is a direct consequence of the "sewing condition" and the identity for the closed string $\delta[12] = -\delta[21]$ where $\delta[12] \equiv \Pi_{\sigma=-\pi}^{\pi} \delta(x_\mu^1(\sigma) - x_\mu^2(-\sigma)) \delta(c^1(\sigma) - c^2(-\sigma)) \delta(\tilde{c}^1(\sigma) + \tilde{c}^2(-\sigma)) = \Pi_{\sigma=0}^{\pi} \delta(x_{e\mu}^1 - x_{e\mu}^2) \delta(x_{o\mu}^1 + x_{0\mu}^2) \delta(c_e^1 - c_e^2) \delta(c_o^1 + c_o^2) \delta(\tilde{c}_e^1 + \tilde{c}_e^2) \delta(\tilde{c}_o^1 - \tilde{c}_o^2)$ and $x_{e(o)}(\sigma) \equiv (x(\sigma) + (-)x(-\sigma))$. Since there is one additional mode in the expansion of c_e , i.e. the zero mode, than there is for $\tilde{c}_o$ we obtain the above identity.

(iv) the **"Leibnitz rule"**-

$$-Q(A * B) = QA * B + (-1)^A A * QB$$

Although this rule appears quite unusual it is nevertheless equivalent to the simple condition on K: $\sum_{i=1}^{3} Q_i K[312] = 0$. Once again the strange sign in this axiom is a direct consequence of the odd Grassmanality of K.

5 SUMMARY AND CONCLUSIONS

We have explored the gauge invariant string field theory introduced by Witten. We have tried to find a representation of the axioms keeping the anti-commuting coordinates. In particular we were able to find an explicit representation of the integral $\int$ for both the open and closed strings. Unfortunately our representation of the $*$ product was incomplete due to the special problems associated with the point $\sigma = \pi/2$ for the open string and $\sigma = 0$ and π for the closed string. We feel however that this technical problem can certainly be overcome [8]. We chose to work with the anti-commuting coordinates in order to be able to use the intuition gained from the component expansions, as described in the text. We believe however that our representation of $\int$ and $*$ is nevertheless equivalent to the path integral representation discussed by Witten. A proof of this assertion would require an explicit representation of the kernel K. One would then be able to explicitly check whether the axioms are satisfied.

In the case of the open string we have assumed that $Q, \int$ and $*$ satisfy the requisite axioms. We have then used these axioms to present the BRST invariant action for the interacting string.

In the case of the closed string we have found the free action and demonstrated that the equations of motion for the massless sector includes the equations of linearized gravity , as well as the equations for the Kalb-Ramond gauge field and a dilaton. We have then obtained an abstract set of axioms which can be used to construct the gauge invariant action for the interacting closed bosonic string. These axioms, though new, are motivated by properties satisfied by the path integral representation of $\int$ and $*$ discussed by Witten. We have assumed that $\int$ has ghost number -3 and $*$ has ghost number +3. The "sewing condition", which is satisfied by the path integral representation , was used to motivate many of the new axioms. In particular, we used the "sewing condition" to prove the important result that the kernel K for the closed string was Grassman odd. Combining this result with Witten's result $\sum_{i=1}^{3} Q_i K[312] = 0$ we obtain the new "Leibnitz rule". There is one property of the $*$ product which was crucial but not sufficiently motivated. We have assumed that $\partial \tilde{c}(A * B) \neq 0$ when $\partial \tilde{c} A = \partial \tilde{c} B = 0$. This relation is probably satisfied by the path integral representation. The two dimensional manifold on which the $*$ product is defined violates ghost number by +3. By the Riemann-Roch index theorem there are 3 nontrivial ghost zero modes on this two dimensional surface. Thus $\hat{c}$ and $\bar{\hat{c}}$ are emitted by the manifold. $\tilde{c}$ is contained in $\hat{c}$ and $\bar{\hat{c}}$ and is probably also emitted. As a result we find $\partial_{\tilde{c}}(A * B) \neq 0$.

If our axioms for the closed string are indeed satisfied by the path integral representation of Witten, we should be able to obtain the scattering amplitudes of the Virasoro-Shapiro model using our action. This calculation should be done.

We are not satisfied with the constraints $\Delta NA = 0$ and $A = \partial_{\tilde{c}} \chi$ which we found necessary to obtain the interacting closed string action. Perhaps these should be thought of as gauge conditions in a theory with an enlarged symmetry group.

These constraints are even more unappealing in the context of the closed superstring as discussed by Banks et al. [13].

Finally, work is now in progress to understand the symmetry principle embodied in the transformation laws of equations (1.20) for the open string and (2.21) for the closed string.

References

[1] W. Siegel, Phys. Lett. 151B , 391; 151B , 396 (1985).

[2] W.Siegel and B. Zwiebach, Nucl. Phys. B263, 105 (1986).

[3] E. Witten , "Non-commutative Geometry and String Field Theory", Princeton preprint (1985).

[4] T. Banks and M.E. Peskin, Nucl. Phys. B264, 513 (1986).

[5] A. Neveu, H. Nicolai, and P. West, Nucl. Phys. B264, 573 (1986); Phys. Lett. 167B, 307 (1986).

[6] N. Ohta, Phys. Rev. Lett. 56, 440 (1986).

[7] S.B. Giddings, "The Veneziano Amplitude from Interacting String Field Theory", Princeton University preprint (1986).

[8] N-P Chang, H-Y Guo, Z. Qiu and K. Wu, "Interacting String Field Theory and Chern-Simons Form", City College preprint CCNY-HEP-86/5 (1986).

[9] A. Jevicki, "Covariant String Theory Feynman Amplitudes", CERN preprint CERN-TH-4341/85 (1985).

[10] H. Aratyn and A.H. Zimerman, "Ghosts and the Physical Modes in the Covariant Free String Field Theory", Hebrew University preprint RIP-85-9 (1985); "Gauge Invariance of the Bosonic Free Field String Theory", Hebrew University preprint Print-85-874 (1985).

[11] H. Hata, K. Itoh, T. Kugo, H. Kunitomo and K. Ogawa, "Manifestly Covariant Field Theory of Interacting String 2.", Kyoto University preprint KUNS-814 (1985); "Manifestly Covariant Field Theory of Interacting String", Kyoto University preprint RIFP-637 (1985).

[12] B. Zwiebach, Gauge Invariant String Actions, MIT preprint MIT-CPT-1308 (1985).

[13] T. Banks, M.E. Peskin, C.R. Preitschopf, D. Friedan and E. Martinec, "All Free String Theories are Theories of Forms", SLAC preprint SLAC-PUB-3853 (1985).

[14] D. Friedan, Phys. Lett. 162B, 102 (1985); "String Field Theory", Chicago University preprint EFI 85-27 (1985).

[15] P. Ramond, "A Pedestrian Approach to Covariant String Theory", University of Florida preprint UFTP-85-18 (1985).

[16] M. Kaku, "Introduction to the Field Theory of Strings", City College preprint CCNY-HEP-85-11 (1985); "Supergauge Field Theory of Covariant Heterotic Strings", Osaka University preprint OU-HET-79 (1985); Phys. Lett. 162B, 97 (1985).

[17] M. Kaku and J. Lykken, "Supergauge Field Theory of Superstrings", City College preprint Print-85-412 (1985).

[18] K. Itoh, T. Kugo, H. Kunitomo and H. Ooguri, "Gauge Invariant Local Action of String Field from BRS Formalism", Kyoto University preprint KUNS 800 HE(TH) 85/04 (1985).

[19] J.-L. Gervais, "Group Theoretic Approach to the String Field Theory Action" , Ecole Normale Superieure preprint LPTENS 85/35 (1985).

[20] K. Bardakci, "Covariant Gauge Theory of Strings", Berkeley preprint UCB-PTH-85/33 (1985).

[21] S. Raby, R. Slansky and G. West, "Toward a Covariant String Theory", Los Alamos preprint LA-UR-85-3794 (1985).

[22] A. Cohen, G. Moore and J. Polchinski, "An Invariant String Propagator", University of Texas preprint UTTG-26-85 (1985); "An Off-shell Propagator for String Theory", Harvard University preprint HUTP-85/A058 (1985).

[23] A. A. Tseytlin, "Covariant String Field Theory and Effective Action",Lebedev Institute preprint LEBEDEV-85-265 (1985).

[24] A. Neveu and P.C. West, "The Interacting Gauge Covariant Bosonic String", CERN preprint CERN-TH-4315/85 (1985).

[25] M.A. Awada, "The Gauge Covariant Formulation of Interacting Strings and Superstrings", Cambridge University preprint Print-85-0937 (1985).

[26] D. Pfeffer, P. Ramond and V. Rogers, "Gauge Invariant Field Theory of Free Strings", University of Florida preprint UFTP-85-19 (1985).

[27] D. Friedan, E. Martinec and S. Shenker, "Conformal Invariance, Supersymmetry and String Theory", Chicago University preprint Print-86-0024 (1985); Phys. Lett. 160B, 55 (1985).

[28] M.E. Peskin and C.B. Thorn, "Equivalence of the Light Cone Formulation and the Gauge Invariant Formulation of String Dynamics", SLAC preprint SLAC-PUB-3801 (1985).

[29] A. Connes , "Introduction to Non-Commutative Differential Geometry", " The Chern Character in K Homology", IHES preprints (1982), to appear in IHES Publication.

[30] C. Becchi, A. Rouet and R. Stora, Phys. Lett. 52B, 344 (1974); I.V. Tyutin, "Gauge Invariance in Field Theory and in Statistical Mechanics in the Operator Formalism", Lebedev preprint FIAN No. 39 (1975).

[31] E.S. Fradkin and G.A. Vilkovisky, Phys. Lett. 55B, 224 (1975).

[32] I.A. Batalin and G.A. Vilkovisky, Phys. Lett. 69B, 309 (1977).

[33] T. Kugo and I. Ojima, Phys. Lett. 73B, 459 (1978).

[34] J. Schwarz, "Fadeev-Popov Ghosts and BRS Symmetry in String Theories", California Institute of Technology preprint CALT-68-1304 (1985).

[35] M. Kato and K. Ogawa, Nucl. Phys. B212, 443 (1983).

[36] S. Hwang, Phys. Rev. D28, 2614 (1983).

BLACK HOLES AND HORIZONS--THE GEOMETRY OF KRUSKAL SPACE AND RINDLER SPACE

Wolfgang Rindler

Department of Physics
University of Texas at Dallas
Richardson, TX 75083-0688

INTRODUCTION

The purpose of these lectures is to provide a simple review of relativity, and in particular to exhibit certain situations in general relativity (GR) whose geometry (especially curvature and topology) are prominent, so as to counterbalance the approach to general relativity which treats gravity as a mere field theory with spin two. This latter approach plays down some of general relativity's most characteristic features. [For some of the basic facts assumed here without proof the reader can consult my Essential Relativity, revised second edition, Springer-Verlag, New York, 1980.]

We begin by reminding the reader of special relativity (SR) whose theme is the embedding of all physics in Minkowski Spacetime $\mathbb{M}$. In "standard" coordinates--namely, orthogonal Cartesian space coordinates x,y,z relative to any inertial frame I, and a time coordinate t read by stationary standard clocks in I synchronized by exchanging light signals--the metric of $\mathbb{M}$ is

$$\underline{ds}^2 = c^2dt^2-dx^2-dy^2-dz^2. \tag{1}$$

We regard $\underline{ds}^2$ as the square of the 4-vector $\underline{ds}$ = (dx,dy,dz,dt) rather than that of its magnitude ds; the former can be positive, negative or zero (corresponding to "timelike," "spacelike," or "lightlike" = "null" displacements $\underline{ds}$, respectively) while the latter is nonnegative. Free particles follow timelike straight lines (geodesics) in $\mathbb{M}$, light rays follow null geodesics, and the proper time elapsed at an arbitrarily moving ideal clock is $\frac{1}{c}\int ds$. It should be noted that this last statement has been borne out to very high accuracy by experience (lifetime of mesons in storage rings, etc.).

The coordinate transformations preserving the RHS of (1) are the general Lorentz (i.e. Poincaré) transformations plus space and time reflections, which we exclude for physical reasons. Physically, Lorentz transformations take us from one inertial reference frame to another in such a way that a signal with speed c in one becomes a signal with speed c in the other as is required by Einstein's relativity principle and the Michelson-Morley experiment. In order for a physical theory to share the symmetries of Minkowski space as required by the relativity principle, its concepts and laws must therefore be Lorenz-invariant. And it is the programme of SR to express all of physics in such form. Maxwell's theory (with a certain transformation law for the field) already possessed the required symmetry, but the nongravitational part of Newton's mechanics needed to be modified, resulting in "relativistic" (Lorentz-invariant) mechanics, with its increase of mass with velocity and $E=mc^2$. The equally necessary modification of Newton's theory of gravity, however, led beyond SR.

Galileo's Principle (the relevant experiment for which consisted in dropping different objects jointly from the top of the Leaning Tower of Pisa) asserts that all particles--independently of their mass and constitution--fall side by side through a gravitational field. Newton's theory accounts for this: acceleration = force x inertial mass, force = gravitational field x gravitational mass, and inertial mass = gravitational mass; hence, acceleration = field. Since inertia and gravity seem to be quite unrelated, the equality of the two kinds of mass was a profound mystery in Newton's theory.

According to the above reasoning, a gravitational orbit is determined by an initial point and velocity (dx/dt, dy/dt, dz/dt). The latter is equivalent to a direction (dx: dy: dz:dt) in the spacetime of events (x,y, z,t). On the other hand it is known that geodesic ("straightest possible") lines in (generally curved) Riemannian spaces are determined by an initial point and direction. Hence Einstein's conjecture, which is the basis of General Relativity (GR): spacetime (though locally Minkowskian) is a curved Riemannian space, its curvature caused by the massive bodies in it according to the field equations; and free test particles and photons follow geodesics in it.

Of course, once spacetime is generalized to accommodate gravity, all the rest of physics must be re-embedded in this newly curved spacetime. Special Relativity, which is the physics of flat spacetime, retains only a local validity, much as plane Euclidean geometry is only locally valid on a curved surface. The link between GR and SR is provided by the heuristic Equivalence Principle (EP). It is known that every Riemannian space, i.e. one that has a homogeneous quadratic metric

$$\underline{ds}^2 = g_{\mu\nu} dx^\mu dx^\nu$$

(Einstein's summation convention!) can be referred to special coordinates in the neighborhood of any given point P such that at P the derivatives $\partial g_{\mu\nu}/\partial x^\sigma$ vanish. Any two such "geodesic" coordinate systems are linearly related at P: $\partial^2 \bar{x}^\mu/\partial x^\nu \partial x^\sigma = 0$ and conversely. If made orthonormal at P, these coordinates correspond to Euclidean coordinates at P. In terms of them, Euclidean geometry is valid locally at P. In particular, in terms of them SR is locally valid at a given event P of spacetime. The generalization of local (e.g. differential) SR laws to GR is achieved by writing down a "covariant" (i.e. tensor) law which reduces to the SR law in geodesic coordinates. This is the mathematical content of the EP. The physical interpretation of the above mathematical result is that around each event P a nearly rigid reference frame S can be found whose fixed points have geodesic worldlines and are therefore freely falling, and whose time slices are orthogonal to those geodesics near P, so that S is also nonrotating and thus locally inertial. The EP then asserts physically that SR holds in any "sufficiently small" freely falling nonrotating orthonormal reference frame. The domain of sufficient accuracy for SR is thus restricted to any freely falling nonrotating lab (local inertial frame) which is small compared to its curvature.

Elapsed time on an ideal clock is still $1/c\int ds$. Also the assumption is made that acceleration does not affect local measurements, i.e. that two observers having the same instantaneous velocity make the same local measurements.

Now consider a static mass distribution with Newtonian potential ϕ in ordinary (Euclidean) 3-space. To show the feasibility of GR, we shall show that to first approximation the Newtonian orbits correspond to the geodesics $\delta\int ds = 0$ of the Riemannian metric

$$\underline{ds}^2 = (1+\frac{2\phi}{c^2})c^2dt^2 - dx^2 - dy^2 - dz^2 \ . \qquad (2)$$

To see this, consider

$$\int ds = \int \frac{ds}{dt}\,dt = \int (c^2 + 2\phi - v^2)^{\frac{1}{2}}\,dt$$

where $v = (dx^2+dy^2+dz^2)^{\frac{1}{2}}/dt$ is the velocity of a freely moving test particle. If ϕ/c^2 and v^2/c^2 are both small compared to unity, we have

$$(c^2+2\phi-v^2)^{\frac{1}{2}} = c[1+(2\phi-v^2)/c^2]^{\frac{1}{2}} \approx c[1+(\phi-1/2v^2)/c^2],$$

and thus, for events P_1, P_2 at times t_1, t_2

$$\int_{P_1}^{P_2} ds = 1/c\int_{t_1}^{t_2} (c^2+\phi-1/2\ v^2)dt = 1/c\int_{t_1}^{t_2} (U-T)dt,$$

where we have written $U= c^2+ \phi$ and $T= \frac{1}{2}v^2$. But the requirement that the last integral be stationary is essentially Hamilton's Principle-obeyed by all Newtonian orbits--and so our assertion is proved.

STATIC SPACETIMES, IN PARTICULAR SCHWARZSCHILD SPACETIME

The simplest spacetimes are static spacetimes, defined as follows: there is to be a spatial framework of "fixed" points at constant radar distance from each other, and the round-trip time for a triangular light signal in that frame is to be constant and the same in both senses. (If the two senses give constant but not necessarily equal times, we call the metric stationary.) These assumptions imply (i) the existence of a unique clock synchronization (up to constant scale and zero-point changes) such that coordinate clocks attached to the fixed points indicate constant multiples of proper time (time indicated by a standard clock), and (ii) a metric of the form

$$\underline{ds}^2 = g_{oo}\ dt^2 - g_{ij}dx^i dx^j \ , \qquad (3)$$

where fixed points have fixed $x^i(i=1,2,3)$, the g's are independent of t, and $(g_{ij}dx^i dx^j)^{\frac{1}{2}}$ gives spatial distance in the framework. If we write

$$g_{oo} = e^{2\phi(x^i)} \qquad (3)$$

in units such that c=1, ϕ is the analogue of Newton's potential in the sense that the proper 3-acceleration of each fixed point is given (exactly) by grad ϕ; thus -grad ϕ can be interpreted as the gravitational field, i.e. as the initial proper acceleration of a particle let go from rest in the framework.

A simple argument shows that the frequency shift between two points P, P'in the framework is given by

$$\frac{\nu}{\nu'} = \sqrt{\frac{g'_{oo}}{g_{oo}}} \ .$$

For since the coordinate running time Δt for successive signals is the same, the coordinate differences dt at the end points are the same, and so

$$\nu/\nu' = ds'/ds = (g'_{00}/g_{00})^{\frac{1}{2}} \quad .$$

Einstein's field equations (in units such that $c = G = 1$)

$$G_{\mu\nu} = 8\pi T_{\mu\nu}$$

relate the spacetime geometry with the sources, $G_{\mu\nu}$ being the Einstein tensor descriptive of the curvature, and $T_{\mu\nu}$ the energy tensor descriptive of the sources. These latter include matter as well as the Maxwell and other fields but not the gravitational field itself: the "gravity of gravity" is taken care of by the nonlinearity of the field equations.

As reference to (3) shows, any static and isotropic spacetime must have a metric of the form

$$g_{00}(r)dt^2 - g_{11}(r)dr^2 - r^2(d\theta^2 + \sin^2\theta d\phi^2)$$

if we adopt polar coordinates with the reference spheres coordinatized in the usual way; of course r no longer necessarily measures radial distance. Applying Einstein's vacuum field equations ($G_{\mu\nu} = 0$) to this metric yields the so-called Schwarzschild metric

$$\underline{ds}^2 = (1 - \frac{2m}{r})dt^2 - (1 - \frac{2m}{r})^{-1}dr^2 - r^2(d\theta^2 + \sin^2\theta d\phi^2), \qquad (5)$$

where, at first, m is merely a constant of integration. But comparison with (2) identifies m as the value of the central mass. According to "Birkhoff's Theorem" the outside field of a spherical mass distribution is given by the static metric (5) even if that distribution pulsates or otherwise changes without violating spherical symmetry.

The argument following (4) then gives the gravitational field as

$$g = \frac{m}{r^2}(1 - \frac{2m}{r})^{-\frac{1}{2}}$$

which differs appreciably from the Newtonian form m/r^2 near the "horizon" $r=2m$, where in fact g becomes infinite. If g is integrated with respect to radial distance from an arbitrary point down to the horizon, the result is 1, or c^2 in full units. Thus, when a test mass M is lowered quasistatically down to the horizon, its total energy Mc^2--including its thermal energy--can be recovered, which would seem to violate the second law of thermodynamics. (Wald and Unruh have addressed and resolved this paradox).

The spherical horizon $r=2m$ is a singularity of the metric (5), though, as we shall see, it is merely a "coordinate" singularity: a transformation of coordinates can make the metric regular there. [Consider, as an analogue, the two metrics $(3\zeta)^{-4/3}\,d\zeta^2 + dy^2$ and $dx^2 + dy^2$ related by $\zeta = (1/3)x^3$; the first has a coordinate singularity at $\zeta = 0$.]

Even though not intrinsically singular (like a locus of infinite curvature) the horizon is of considerable physical interest. It represents a spherical light front of constant surface area--running in an outward or inward direction. How can that be? First note that while (5) satisfies Einstein's vacuum field equations for $r > 2m$ and $r < 2m$, the "inner" region $r < 2m$ is not static. There the only positive term in the metric is the second, and thus $r = \text{const}$ implies $\underline{ds}^2 < 0$ which is impossible for a particle (or photon) worldline. Thus, for a particle, r must change; r is therefore a time coordinate. And for the sake of causality we must make a choice: either (i) r increases for all particles or (ii) r decreases for all particles, into the future. (A similar choice must of course be made for t in the outside region.) If the choice (i) is made, the sphere r = 2m is an inward pointing light front, if (ii), outward, as we shall see more clearly later.

But let us first examine the spatial geometry of the outside Schwarzchild region r > 2m. It is determined by the three-dimensional metric

$$d\sigma_3{}^2 = (1 - \frac{2m}{r})^{-1} dr^2 + r^2(d\theta^2 + \sin^2\theta d\phi^2), \tag{6}$$

which in turn can be represented by its typical cut $\theta = \frac{\pi}{2}$ with corresponding metric

$$d\sigma_2{}^2 = (1 - \frac{2m}{r})^{-1} dr^2 + r^2 d\phi^2. \tag{7}$$

For the distance between two circles $r = r_1$, $r = r_2$, in (7) is identical to the distance between the full spheres $r = r_1$, $r = r_2$ in (6). But (7) can easily be shown to be the metric of "Flamm's Paraboloid" of revolution, whose equation in cylindrical polar coordinates is $Z^2 = 8m(r-2m)$ (see Fig. 1). Since the slope of a parabola tends to zero, the top half of the paraboloid is to all intents and purposes a plane with a beveled hole in the middle (see Fig. 2).

Its Gaussian curvature $-m/r^3$ falls off very quickly with r from a maximum of $-1/8m^2$ at the waist r = 2m. We may note that in the present units which make c = G = 1, m has the dimensions of a length. For example, m_{sun} = 1.5km and m_{earth} = 0.5cm. Thus, in general, the waist r = 2m (horizon) lies far inside the central mass, where the vacuum metric (5) no longer applies.

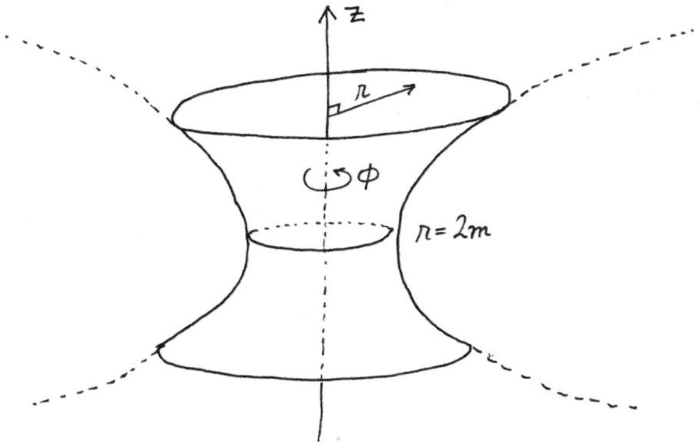

Fig. 1

Fig. 2

An r,t diagram of the light cones in Schwarzschild space (5) also teaches us many things. Note that, under suppression of the angular coordinates, the light-cones (the bundle of directions satisfying $\underline{ds}^2=0$) are given by

$$dt = \pm \frac{dr}{1 - \frac{2m}{r}} \quad . \tag{8}$$

Figure 3 shows some typical light cones at t = 0 (they are independent of t), and for the sake of visualization they are shown 3-dimensional: particle paths must lie inside them and light paths along their mantle. To get the r,t equation of radial light signals we integrate (8):

$$\int dt = \pm \int \frac{dr}{1-2m/r} = \pm \int [1 + \frac{2m}{r-2m}]\, dr,$$

and this shows that $t \to \pm\infty$ as $r \to 2m$. It therefore takes an infinite coordinate time to reach the horizon--but that does not mean the horizon will not be reached, or crossed! The above-mentioned coordinate transformation which makes the horizon regular makes it plain (and we shall see this below) that particles can cross the horizon at finite proper time and photons at finite "affine parameter." One simply must accept that any such crossing always occurs at $t=\pm\infty$ in Schwarzschild time.

Fix attention on the radial light signal A in Fig. 3. It has t↓ (shorthand for: decreasing into the future), but that is unimportant since r is here the time; and we can suppose that we are in an inside region with r↓ for all particles. A is evidently the continuation of some outside worldline like A'. Now B is as good a light signal as A in the same region. It is the continuation of B' but that has t↓. So A' and B' cannot coexist in the same outside region, since t here __is__ a time. To serve as "feeders" for

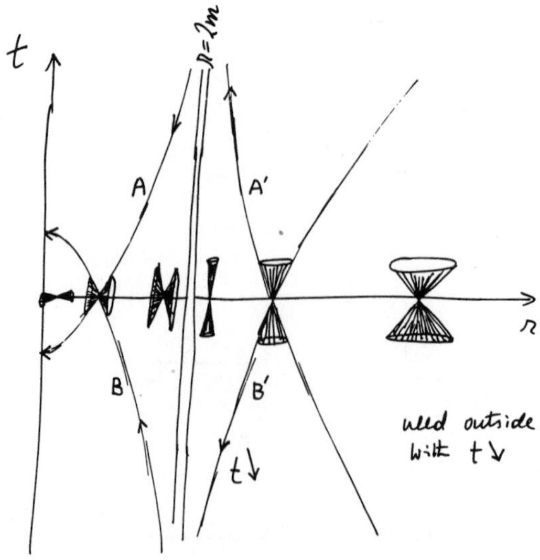

Fig. 3

an inside region with $r\downarrow$ we therefore need two separate outside regions, one with $t\uparrow$, one with $t\downarrow$. However, in the $t\uparrow$ outside region, -B' (B' traversed in the opposite sense) is a possible light path; what is its continuation beyond the horizon? Evidently -B! So as feeder for the $t\uparrow$ outside region we need an inside region with $r\uparrow$. For all geodesics to have continuations we thus need to patch two outside regions (we call them I and III) to two inside regions (II and IV) as shown in Fig. 4. The paths labeled A,A', B,B' in Fig. 4 correspond to those labeled by the same letters in Fig. 3. The path C, C', C" corresponds to a particle: it comes out of IV, rises and falls back in I, but when it crosses the horizon for a second time it must fall into a different inner region namely one with $r\downarrow$. How are we to understand this strange topology? Krushal has shown how all four regions I, II, III, IV fit neatly into a "maximally extended" spacetime, now called Kruskal space. But before we discuss this, we make a lengthy digression back to special relativity in order to prepare the ground.

THE SPECIAL-RELATIVISTIC UNIFORMLY ACCELERATED ROCKET

Consider a particle moving along the x-axis of an inertial frame S such that at every instant its proper acceleration (i.e. its acceleration relative to its instantaneous rest frame) is the same, say α. In units such that c=1 its motion can be described by the equation

$$x^2 - t^2 = \alpha^{-2} \quad , \tag{9}$$

of which Fig. 5 (a) is a graph. This graph is a hyperbola, whence such motion is called "hyperbolic." Figure 5 (b) for comparison shows a Newtonian worldline with constant acceleration α: this is "parabolic."

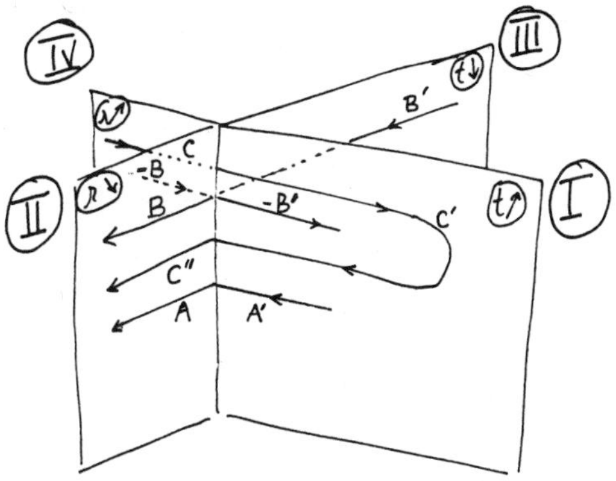

Fig. 4

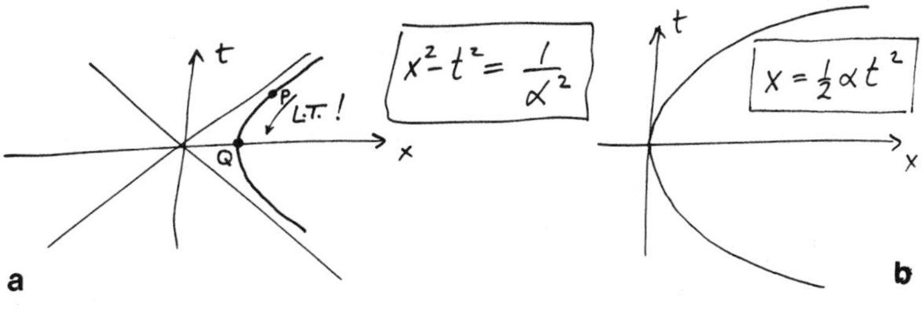

Fig. 5

The simplest way to see that (9) indeed represents motion with constant proper acceleration α is to apply an active Lorentz transformation to Fig. 5 (a), such that a given event P maps into the previous vertex event Q: this corresponds to looking at the situation from the rest frame at P. Eq. (9) goes into itself, so the hyperbolic graph remains unchanged, only its various points slide downwards. By this argument it is seen that the proper acceleration at any event P is the same as that at Q, where $d^2x/dt^2 = \alpha$, from (9). Now consider a long rocket moving along the same x-axis. Suppose its front moves hyperbolically, with constant proper acceleration α. The back at each instant moves faster, for not only must it keep up with the front, but in addition the rocket's length diminishes owing to length contraction. There is a certain kind of motion called Born-rigid motion, characterized by each volume element of a moving body preserving its dimensions in its successive rest-frames. Clearly, this is necessary and sufficient for no internal stresses to develop. As seen from a single inertial frame used for the description of the entire motion, at each instant each volume element is contracted in the direction of its velocity by the Lorentz factor corresponding to that velocity. (See Fig. 6 which shows three intrinsically equal volume elements.)

It comes as a pleasant surprise that if one point of a Born-rigidly moving rocket moves hyperbolically, all of its points do, though with progressively higher constant proper acceleration towards the rear. In fact, at a certain point the acceleration becomes infinite, and the rocket cannot be continued beyond that. A Minkowski diagram of this situation makes everything clear. Figure 7 exhibits several members of the family of hyperbolic worldlines

$$x^2 - t^2 = X^2 \qquad (10)$$

parametrized by X which together constitutes one "fiber" of the rocket. X measures the reciprocal of the proper acceleration of each worldline, and also their proper distance apart at t=0 (and, as we shall see, always).

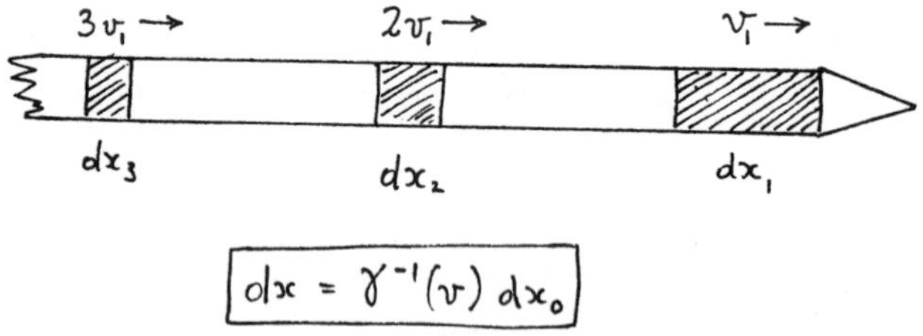

Fig. 6

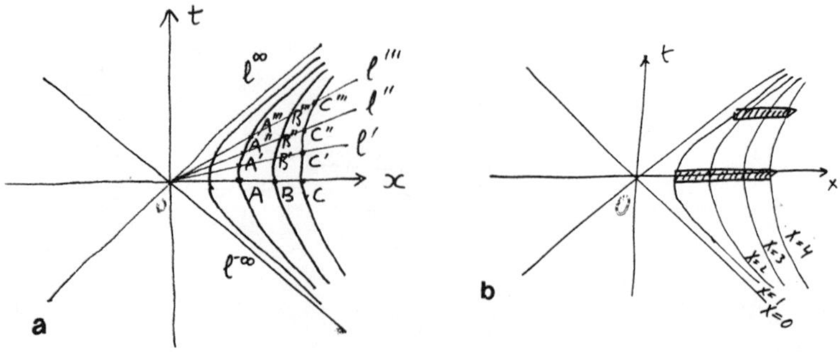

Fig. 7

Consider a sequence of lines ℓ', ℓ'', ℓ''',... through 0 in Fig. 7(a) as shown; each of these is the x-axis of an inertial frame. By Lorentz transformations the x-axis of our original frame S can be successively transformed into ℓ', ℓ'', ℓ''',... with $A \to A' \to A'' \to A''' \to \ldots$ and similarly for any other points B, C, ... In each of these successive inertial frames the rocket is therefore instantaneously at rest, with the same distances between its various points as in S. This establishes that it moves Born-rigidly. Figure 7(b) shows successive instantaneous views of the rocket in the base frame S. Note its progressive Lorentz contraction relative to S.

Since an observer cannot distinguish intrinsically between traveling at proper acceleration α and being at rest in a gravitational field $-\alpha$, the rocket gives its inhabitants the illusion of being a skyscraper at rest in a variable gravitational field of magnitude 1/X (see Fig. 8). Note that the common asymptotes $\ell^{\pm\infty}$ (corresponding to X=0) of all the hyperbolic worldlines in Fig. 7(a) represent light signals. In each of the successive rest frames of the rocket these signals (or photons) are at the same distance from the back end (e.g. 1 unit from X=1) of the rocket. Though the rocket can be extended only to an arbitrarily small but finite value of X, we may loosely speak of the "bottom floor" of the rocket-skyscraper as made up of photons. At time t=0 ("halfway through eternity" if the rocket exists forever) one bottom floor (ℓ^{∞}) peels off, to be replaced by another ($\ell^{-\infty}$).

MINKOWSKI SPACETIME AS SEEN FROM TWO SPACE-FILLING ROCKETS (RINDLER SPACE)

Consider once more Minkowski space $\mathbb{M}$. Instead of a single "line" rocket along the positive x-axis, imagine infinitely many identical such fibers side by side, moving as a block according to eq. (10) and filling the entire region $-x < t < x$, $-\infty < y < \infty$, $-\infty < z < \infty$. Add to this its mirror image in the plane x=0. (See Fig. 9). Of course, even these two rockets do not fill

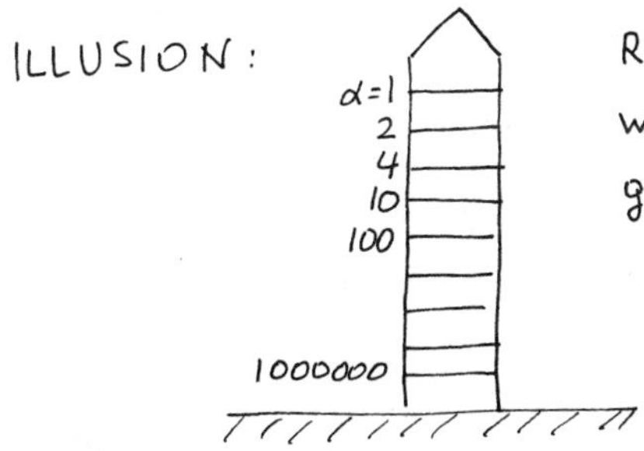

Fig. 8

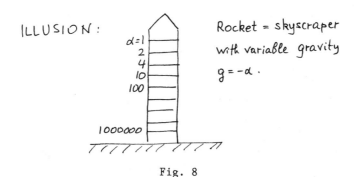

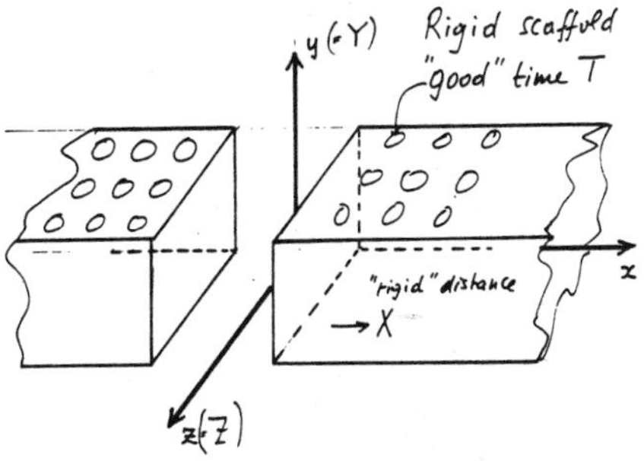

Fig. 9

the entire 3-space: except at t=0 there is a gap between them. But it is of interest to examine how inhabitants of these rockets would most naturally coordinatize and view their rockets and the space between.

Rewrite (10) in terms of yet another parameter T:

$$\left.\begin{array}{l} t = X \sinh T \\ x = X \cosh T \end{array}\right\} \quad x^2 - t^2 = X^2 \tag{11}$$

Then taking c = 1 and also setting

$$y = Y, \; z = Z, \tag{12}$$

we have

$$\underline{ds}^2 = dt^2 - dx^2 - dy^2 - dz^2 = X^2 dT^2 \underbrace{- dX^2 - dY^2 - dZ^2}_{-d\sigma^2} \; . \tag{13}$$

The latter we recognize as the metric of a static gravitational field [cf. (13)] with T as the appropriate time coordinate, and with a rigid frame defined by X,Y,Z = constant [and, incidentally, with a gravitational field 1/X--cf. after (4)]. These coordinates and this metric are "natural" for life in the rockets. X>0 refers to the right, X<0 to the left rocket. The space between can now be coordinatized as above, but with t and x interchanged. (See Fig. 10) This, however, leads to a metric of the form $-X^2dT^2+dX^2-dY^2-dZ^2$ and so it is convenient to make yet another transformation,

$$2R-1 = \pm X^2 \tag{14}$$

(the positive sign applying inside, the negative sign outside the rockets) which results in a uniform metric form throughout $\mathbb{M}$:

$$\underline{ds}^2 = (2R-1)dT^2 - (2R-1)^{-1}dR^2 - dY^2 - dZ^2 \; . \tag{15}$$

Figure 10 shows the relations between the various coordinates.

Since it is our aim to make an analogy between (15) and the Schwarzschild metric, and since the latter has an intrinsic singularity, we shall now give one to the former: let us cut out of $\mathbb{M}$ the two regions R<0 (cf. Fig. 11). This "mutilated" Minkowski space (often referred to as Rindler space) has two singular "edges," E_1 and E_2. They are not curvature singularities but nevertheless they are intrinsic singularities.

We note that between the rockets, i.e., for 0<R<1/2, R=const results in $\underline{ds}^2<0$ so that, as for the Schwarzschild inner region, a choice $R\uparrow$ or $R\downarrow$ must be made. The $R\uparrow$ region corresponds to IV in Figs. 10, 11 while $R\downarrow$ corresponds to II. Similarly, for R>1/2 a choice $T\uparrow$ or $T\downarrow$ must be made: these correspond to regions I and III, respectively. As in the case of Schwarzschild, all four regions are necessary in order to account for the entire length of geodesics, and a diagram like Fig. 4 might be a first attempt by rocket people to understand their world--before they discover Minkowski space.

We shall make an analogy between the "rocket fibers" in $\mathbb{M}$ and radial "skyscrapers" erected in Schwarzschild space around the horizon r=2m (which, as we have stated, is a light front). Of course, these skyscrapers must be mutually braced so as to prevent them from falling in. (See Fig. 12)

Extension to all Mink. Space

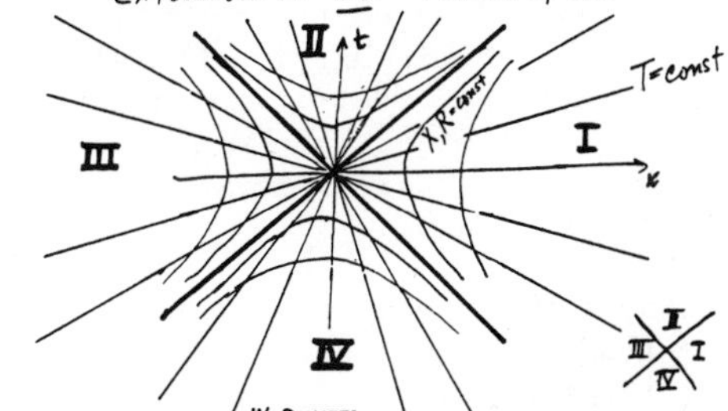

	IN ROCKETS I(X>0) III(X<0)	BETWEEN ROCKETS II(X>0) IV(X<0)
$t=$	$X\sinh T$	$X\cosh T$
$x=$	$X\cosh T$	$X\sinh T$
$t/x=$	$\tanh T$	$\coth T$
$x^2-t^2=$	X^2	$-X^2$
$dt^2-dx^2=$	$X^2dT^2-dX^2$	$-X^2dT^2+dX^2$

$$(2R-1) = x^2-t^2$$

$$\Rightarrow \boxed{ds^2 = (2R-1)\,dT^2-(2R-1)^{-1}dR^2-dY^2-dZ^2}$$

Fig. 10

$$\boxed{ds^2 = (2R-1)\,dT^2-(2R-1)^{-1}dR^2-dY^2-dZ^2}$$

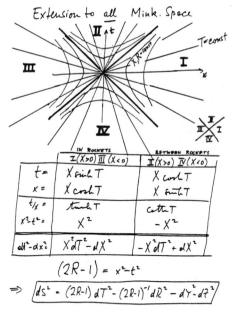

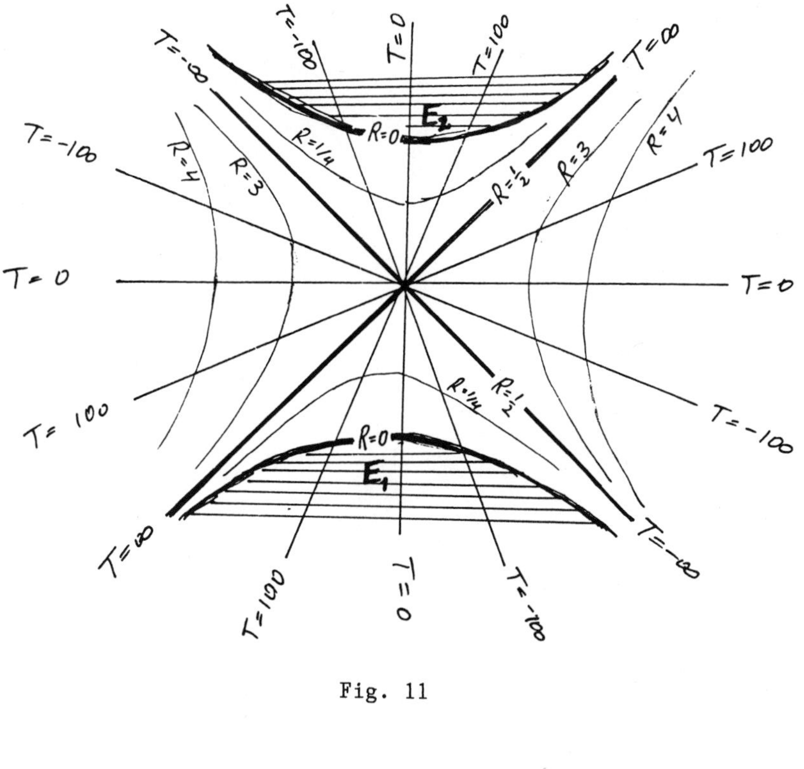

Fig. 11

For the rockets it is clear from Figs. 10, 11 that their "photon basement" is a horizon in the following sense: if the rockets have a central elevator shaft, then an object dropped through the horizon can never be retrieved. In fact, it cannot avoid hitting the singularity E_2 since that now bounds its forward light cone. But if an observer at a fixed level in the rocket observes the falling object, he will see it meet the horizon only at his time $T=\infty$; in other words, he sees the object's history just before it meets the horizon infinitely dilated.

Note also: "Halfway through eternity" (namely at t=0) the "photon basement" changes from one that is penetrable only into the rocket to one that is penetrable only out of the rocket.

The right and the left rockets are causally disconnected: no light signal from one can penetrate into the other. (But of course light from both can be simultaneously observed in region II.)

Every free (straight) worldline in mutilated Minkowski space is finite. It originates on E_1 and ends on E_2 after a finite proper time. Infinite life can be had only by utilizing time dilation and accelerating away from the horizon, for example by traveling in one of the rockets.

The edge E_1 is forever visible to an observer fixed in a rocket say at $R=R_0$. Moreover, as measured by the proper time elapsed on free particles that leave E_1 and arrive at R_0 with zero velocity (i.e. fall back after reaching a "height" R_0), E_1 is at "constant distance" from R_0. This is clear from Fig. 13, where any worldline such as AB can be Lorentz transformed into any other such as CD. As a result, the rocket people in either rocket might mistakenly picture E_1 as a plane dragged at constant distance behind the rocket. (cf. Fig. 14)

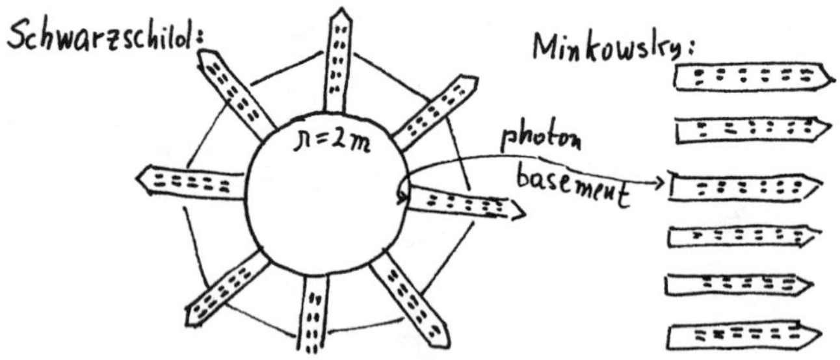

Fig. 12

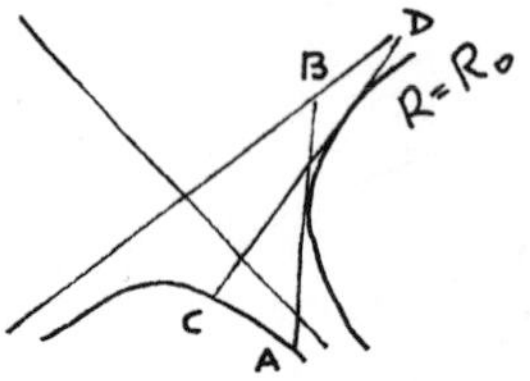

Fig. 13

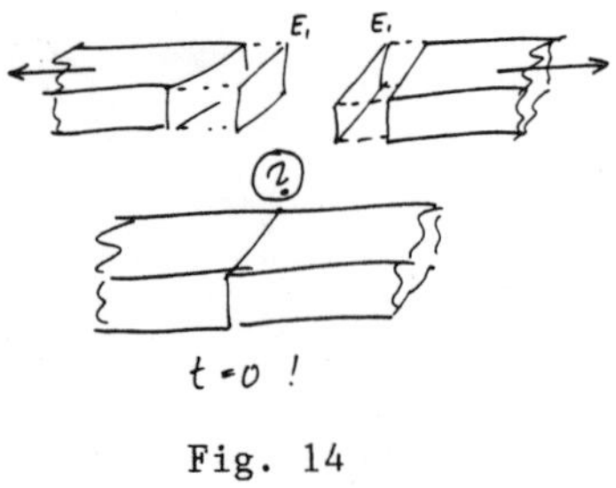

Fig. 14

Moreover, they might think of E_2 in the same way. Yet at t=0 the rockets touch and neither E_1 nor E_2 is between them! For similar reasons people in Schwarzschild skyscrapers quite mistakenly picture the singularity at r=0 as the central "point" of the spherical horizon.

We shall see that, in fact, all the above-mentioned apparent puzzles have relevance also for Schwarzschild space. And just as in the case of the rockets Minkowski coordinates and Minkowski space bring clarity, so Kruskal coordinates and space bring clarity to the Schwarzschild geometry.

KRUSKAL SPACE

Having chosen units to make c=G=1 in the Schwarzschild metric (5), and having length, time and mass units at our disposal, we can use the last freedom to make m=1/4. We shall also now write R and T for the Schwarzschild coordinates r and t in the metric (5), so that it now becomes

$$\underline{ds}^2 = (1 - \frac{1}{2R})dT^2 - (1 - \frac{1}{2R})^{-1}dR^2 - R^2(d\theta^2 + \sin^2\theta d\phi^2). \tag{16}$$

It will be a little while before we actually make use of this metric.

We are now ready to define Kruskal space. It is the manifold corresponding to the metric

$$\underline{ds}^2 = \frac{1}{2Re^{2R}}(dt^2 - dx^2) - R^2(d\theta^2 + \sin^2\theta d\phi^2), \tag{17}$$

in which R is a function of $x^2 - t^2$ given implicitly by

$$e^{2R}(2R-1) = x^2 - t^2, \tag{18}$$

(see Fig. 15) and in which the angular coordinates θ,ϕ range over the usual values while the Kruskal coordinates x and t (elsewhere usually denoted by u and v) are restricted by the condition R>0. R=0 is the only singularity of the metric (17) and indeed it is a locus of infinite curvature. Everywhere else, as we shall see, the metric satisfies the Einstein vacuum field equations. Evidently Kruskal space is spherically symmetric; its main interest lies in the x,t part. The coordinate t is the time coordinate, but since R contains t the metric is not static; x is a radial coordinate. In a Kruskal diagram (or map), with x and t as Cartesian coordinates (see Fig. 16), the lines R = constant are hyperbolas as shown. Each of its points corresponds to an entire 2-sphere of radius R in the full spacetime, coordinatized by the suppressed coordinates θ,ϕ. And all ± 45° lines represent light, just as in a Minkowski diagram. (cf. Fig. 17) So the lines R=1/2 represent spherical light fronts of constant radius!

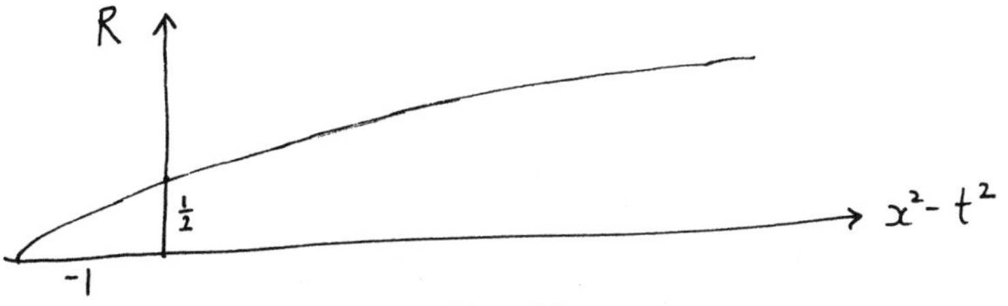

Fig. 15

But before discussing its further properties, we shall show how Kruskal space is, in fact, composed of two outer and two inner Schwarzschild regions. Moreover the relation between the Kruskal coordinates x, t and the Schwarzschild coordinates R, T is almost the same as that between the Minkowski coordinates x, t and the rocket coordinates R, T. The only difference is this:

$$x^2 - t^2 = \begin{cases} 2R-1 & \text{(Minkowski)} \\ e^{2R}(2R-1) & \text{(Kruskal)} \end{cases}$$

If we introduce an auxiliary variable X as in the case of the rockets, the upper table in Fig. 10 applies here also and a final conversion of the X, T metric to R and T now yields quite straightforwardly the Schwarzschild metric (16) in all four quadrants. As expected, I has T $\uparrow$, II has R $\downarrow$, III has T $\downarrow$ and IV has R $\uparrow$. Now since the vacuum field equations are tensorial, and since the inside and outside Schwarzschild metrics satisfy them, Kruskal space must satisfy them everywhere except possibly at R=1/2, where the Schwarzschild metric is singular. But since the Kruskal metric is continuous at R = 1/2, it satisfies the vacuum field equations also there, and thus everywhere in the domain of its coordinates. Incidentally, the locus R=1/2 ("horizon") is now established as intrinsically regular, Kruskal coordinates being only one of many possible systems for regularizing the metric there.

A very important property of the Kruskal metric (17) is its invariance under standard Lorentz transformations in x and t (since these leave both x^2-t^2 and dt^2-dx^2 invariant.) So the Kruskal diagram (Fig. 16) possesses essentially the same invariance properties as the Minkowski diagram. (Fig. 11) Some immediate corollaries follow:

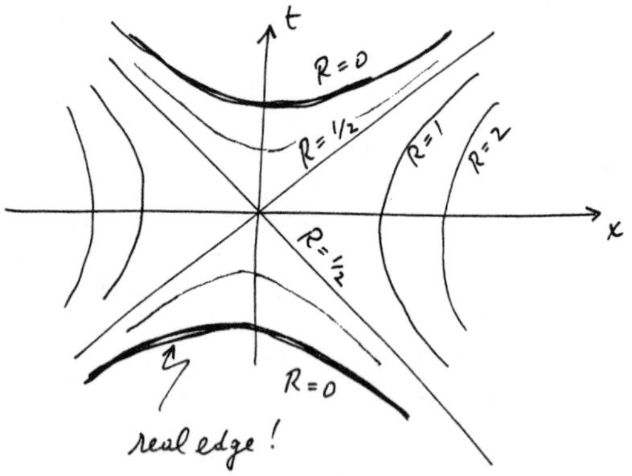

Fig. 16

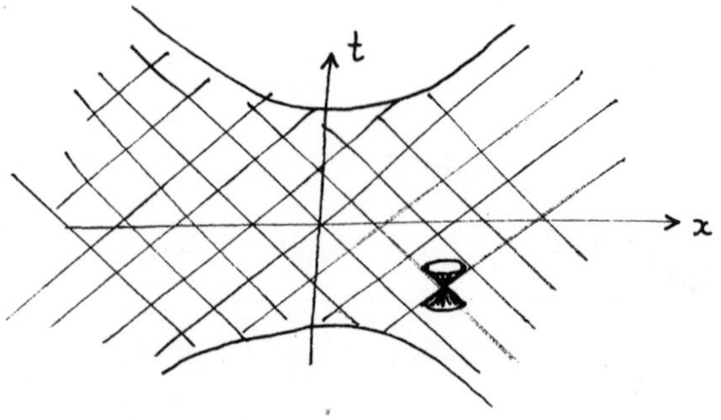

Fig. 17

(i) All spatial sections T = constant of Kruskal space are isometric, since each is Lorentz-transformable to every other.

(ii) All worldlines T,θ,ϕ = constant in quadrants II and IV are timelike geodesics of equal total length $2\int_0^{\frac{1}{2}} (1/2R-1)^{-\frac{1}{2}} dR = \frac{\pi}{2}$, since all are transformable to T = 0, which is obviously a geodesic by symmetry. (As a matter of fact, on a much larger scale than one usually associates with the Schwarzschild metric, this bundle of worldlines can be used to represent the galaxies of a "cylindrical" test-dust model universe of finite proper duration that is homogeneous but not isotropic!)

(iii) All worldlines R = constant > 1/2, θ,ϕ = constant (the hyperbolas in quadrants I and III) have constant proper acceleration, since each event on one of them is transformable to every other.

(iv) The aggregate of lines R = constant > 1/2 in quadrant I (and similarly those in quadrant III) represent a uniformly accelerated rocket fiber moving Born-rigidly in a radial direction through Kruskal space. The proof is precisely as in Minkowski space. These rockets will turn out to be the "skyscrapers" of Schwarzschild space shown in Fig. 12. In the Schwarzschild metric, however, they have "co-moving" coordinates and thus they appear static in a static field. Moreover, the "mirror image" of each rocket, as well as the ever increasing space between these two, must all be sought within the horizon sphere on which each rocket stands! That horizon sphere in the Kruskal diagram corresponds to the boundary of quadrant I, and it is clear that much more goes on "beyond" that boundary than meets the eye of the skyscraper people.

One good way to gain an understanding of the geometry of Kruskal space is to regard it as the time development of a spacelike 3-space, in other words, to slice it into a sequency of spacelike cross-sections. Of course, one has an infinite choice for doing this. Cross-sections of constant Kruskal time t are one possibility, and they have been used by Wheeler and his students. Their disadvantage is that for $|t| > 1$ these sections consist of two disconnected portions, as inspection of Fig. 16 makes clear. An alternative slicing which avoids this unphysical splitting in two of a connected manifold is shown in Fig. 18. The exact definition of the "time" τ which is constant on these slices is less important than its continuous variation from $\tau = 0$ on R = 0 in quadrant IV to a finite final time, say $\tau = \frac{\pi}{2}$, on R = 0 in quadrant II. For definitiveness, we can choose a set of

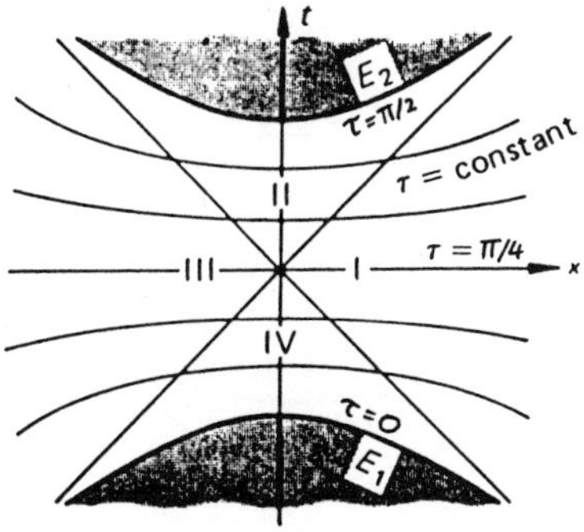

Fig. 18

confocal hyperbolas in the Kruskal diagram and label them continuously from $\tau = 0$ to $\tau = \frac{\pi}{2}$. On this view Kruskal space is a spacetime of finite total duration $\frac{\pi}{2}$, as is indeed natural for the analytic extension of a finite-duration homogeneous universe (in II and IV). The geometry of each slice becomes clear when we remember that each point in the Kruskal diagram represents an entire 2-sphere of radius R. So each slice is generated by a family of 2-spheres, their distance apart being $\int ds$ along the corresponding line in the Kruskal diagram, and their radii varying from infinity at $x = -\infty$ down to a minimum at $x = 0$ and back to infinity at $x = +\infty$. We can again (as in the case of Schwarzschild outer space) represent each such 3-space by a typical 2-dimensional "symmetry" cut $\theta = \frac{\pi}{2}$, reducing 2-spheres to circles, whose normal distance apart corresponds to the normal distance between the spheres in the full 3-space. This sequence of 2-dimensional spaces is shown in Fig. 19, and consists of a nested set of "double trumpets," whose outermost member--representing the line $t = 0$ $(T = 0)$ in the Kruskal diagram--is in fact the Flamm paraboloid of Fig. 1. On each of these double trumpets there are two circles (shown stippled) of special interest, namely, those of radius $R = 1/2$, representing the horizon light fronts. The Kruskal spacetime suddenly appears out of nowhere as an infinite line E_1. This immediately flares open into a long drawn-out double trumpet, reaches maximum girth and maximum flare as a Flamm paraboloid, whereupon the entire sequence is reversed, back to a line E_2 momentarily, and then again nothing. As these 3-spaces open, observe how the horizons rush towards each other at the speed of light. While regions I and III (outside the horizons) grow, region IV (between the horizons) shrinks. As the horizons cross, region IV disappears altogether and II appears. An observer at constant $R = R_0 > 1/2$ is similarly running along the trumpets (picture a circle of radius R_0), but with constant proper acceleration outwards.

The following alternative slicing of Kruskal space brings out even more clearly its relation to Schwarzschild space. Take the same confocal hyperbolas as in Fig. 18, but where they intersect a given hyperbola $R = 1/2 + \varepsilon$ very close to the horizon, continue them with T = constant (heavy line in Fig. 20.) Each such slice consists, essentially, of two Flamm-paraboloid

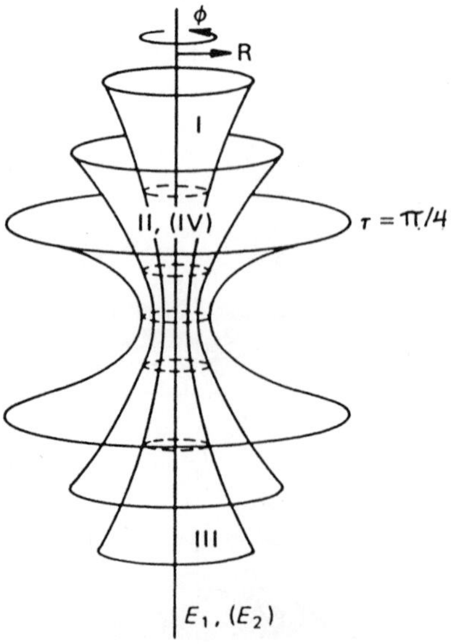

Fig. 19

halves (corresponding to T = constant, $R > 1/2 + \varepsilon$) joined by the portion of the double trumpet of Fig. 18 that lies between the horizons (strictly: between the horizons "plus epsilon"). If we approximate the very flat Flamm-paraboloid halves by planes as in Fig. 2, our sections look as shown in Fig. 21. The planes with their horizon "hole" remain unaltered (static Schwarzschild exterior spaces!), while the "neck" between them changes from infinite length and zero girth to zero length (at which time the horizons cross--this corresponds to the touching of the rockets in Fig. 14), and then a time-reversal of all this. An observer "at rest" (constant $R > 1/2$) in one of the "planes" sees nothing of the expanding phase of the neck. If he manages to stay out of the hole, whose attraction he must resist (with rigid struts to other observers around the hole, or alternatively with a rocket engine) he can live forever. But if he falls in past the horizon, then he meets the singularity in a finite proper times, i.e. he gets squeezed out of existence in the contracting neck. Note that the "opposite" rockets in Kruskal space corresponding to the two rockets of Fig. 14, whose bottom photons get exchanged at t = 0, are shown in Fig. 21 as K_1, K_3; they are not diametrically opposite each other across the horizon in a single exterior Schwarzschild space as are K_1 and K'_1. Figure 21 illustrates nicely how the two outer spaces I and III can momentarily share an entire horizon 2-sphere (a circle in Fig. 21) without ever sharing any other event or signal.

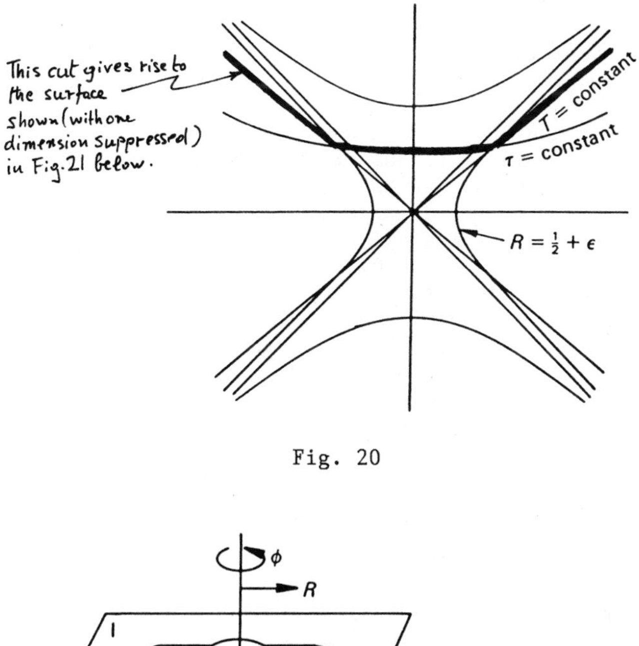
This cut gives rise to the surface shown (with one dimension suppressed) in Fig. 21 below.
T = constant
τ = constant
R = ½ + ε

Fig. 20

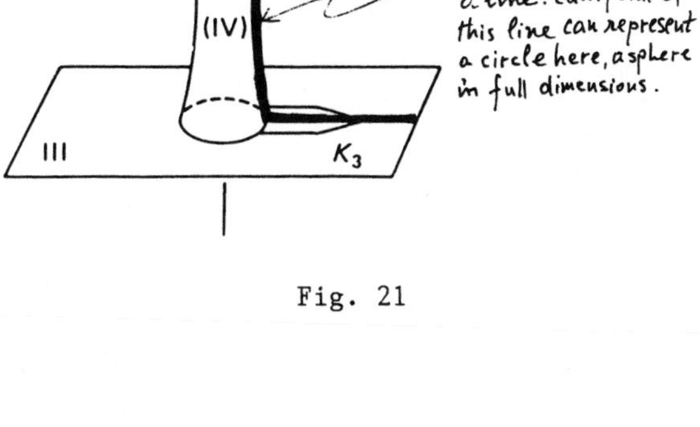
φ
R
I
K'1
K1
II
(IV)
III
K3
The entire Kruskal diagram concerns the history of such a line. Each point of this line can represent a circle here, a sphere in full dimensions.

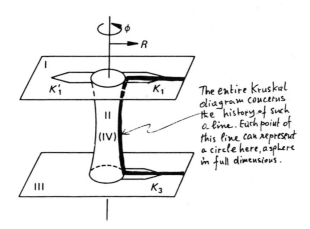

Fig. 21

A possibly fruitful image is the following: as the neck shortens in Fig. 21, "space" rushes out into the planes, which expand radially. The horizon-circle moves inwards over this space at the speed of light (it is a potential light front). During this stage signals and particles can evidently come out of the neck. After the horizons cross and the neck begins to expand again, "space" rushes into the neck from the planes, the horizon-circle moves outwards at the speed of light, and signals and particles can cross into the neck.

The neck as such, namely regions II and IV, can be represented by space-like cuts $R = \text{constant} < 1/2$, if the continuation across the horizons is of no interest. These cuts are the spaces of constant cosmic time $\int(1-1/2R)^{\frac{1}{2}}\, dR$ of the homogeneous cylindrical universe mentioned earlier. Under suppression of one angular coordinate (putting $\theta = \frac{\pi}{2}$), as we did before, these cuts can be represented by infinite cylinders of radius R. The "universe" begins as an infinite line E_1, which immediately opens into a cylinder, which while growing in circumference shrinks longitudinally. At cosmic time $\frac{\pi}{4}$ (corresponding to Kruskal time $t = 0$)--halfway through its existence--the cylinder has shrunk to a single circle (a 2-sphere of radius $R = 1/2$ in full dimensions) and then the whole development repeats in time-reversed order until the universe again becomes a line, E_2, and then nothing.

What is such a cylindrical universe like, momentarily, in the full 3 dimensions? Picture a sequence of nested spheres, each entirely inside the one before, and at the same distance from it. The sequence never stops, in either direction, and all these spheres have the same surface area!

This almost concludes our discussion of full Kruskal space--a "white hole" which becomes a "black hole": For the "first half of eternity" its two horizons are penetrable outwards (white hole), for the second half, inwards (black hole). Its initial singularity E_1 is visible forever in the outside regions I, III. The existence of such a full Kruskal space, suitably merging at its far ends (large R) with the spacetime we live in, would violate an important conjecture on which much of advanced black-hole theory is predicated, namely, "cosmic censorship," according to which no "naked" (i.e. distantly visible) singularities must exist. For these would interfere with the predictability of physics, singularities being places where the known laws of physics do not apply.

John Wheeler and his school at one time hoped to construct a geometric theory of elementary particles, in which full Kruskal spaces together with their "electrically charged" generalizations, and a spacetime honeycombed with Kruskal-type "wormholes" would play a basic role. ("Matter without matter," "charge without charge," "geometry is everything.") However, that beautiful idea eventually ran into insurmountable difficulties and had to be abandoned.

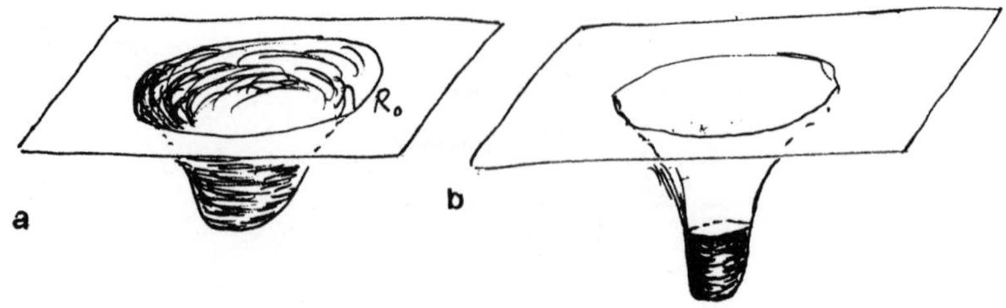

Fig. 22

A Kruskal white hole-black hole would have to be created ab initio: it cannot develop from a collapsing object. For example, the full static spacetime near an ordinary star consists of one single outer Schwarzschild region I up to the star's surface $R = R_0$, and is then continued by a static internal spacetime whose space sections are positively curved (Fig. 22(a). If through gravitational self-attraction the outer surface of the star shrinks to the horizon (while the outer Schwarzschild region continues into the region previously occupied by the star, by Birkhoff's Theorem), its complete collapse in a finite proper time cannot be avoided. Any point on the star's surface then traces out a worldline in the Kruskal region II, and the maximum proper time along such a worldline is $\frac{\pi}{4}$. This can be seen by inspection of the interior Schwarzschild metric (16):

$$\underline{ds}^2 = (\frac{1}{2R} - 1)^{-\frac{1}{2}}dR^2 - (\frac{1}{2R} - 1)dT^2 - R^2(d\theta^2 + \sin^2\theta d\phi^2).$$

Obviously for maximum $\int ds$ (with $\underline{ds}^2 > 0$) we need $dT = d\theta = d\phi = 0$ and this gives $\int_0^{\frac{1}{2}} (\frac{1}{2R} - 1)^{-\frac{1}{2}}dR = \pi/4$. Figure 22(b) shows the collapse in progress: the material of the star closes the neck, which nevertheless elongates and finally ends--with the star--in a curvature singularity. It must be admitted that all we have shown here is that the surface of the star must pinch off to a singular point. Can the interior escape annihilation--e.g. by the blackened region in Fig. 2(b) bulging out into a sphere (hypersphere in full dimensions)? One would need to examine the possible interior solutions in detail, but the answer seems to be "no."

APPENDIX ON HYPERSPHERES

Having mentioned a hypersphere, let us elaborate a little on this. Hyperspheres are of particular interest in isotropic cosmologies, where they provide the constant-density "instantaneous" space sections of closed model universes. A hypersphere S_3 is the 3-dimensional analogue of an ordinary 2-sphere (spherical surface) S_2 and it can be pictured as a set of nested 2-spheres (just as the 2-sphere is a set of nested circles). We start with a point, surround it with a small sphere, and that with a slightly larger one, etc., each at constant radial distance from the one before. As we continue this process, we find that the areas of the spheres at first increase, though at a lesser rate than if we did all this in flat (Euclidean) 3-space; then a sphere of maximum area, say $4\pi a^2$, is reached (the equatorial sphere), whereupon the areas begin to decrease until we reach a last sphere (the antipodal sphere) whose area is zero, and whose radial distance from the origin is πa. The total volume of the hypersphere is $2\pi^2 a^3$. All its points are equivalent, and each can serve as origin in the above description. The geodesics that issue from any one of its points along co-planar directions are all closed lines of length $2\pi a$ and they form a 2-sphere of radius a. The inside and outside of each such sphere ("geodesic plane") are identical halves of S_3, just as the "inside" and "outside" of any great circle on S_2 are identical halves of S_2. If I blow up a rubber balloon in S_3, its surface will increase until it looks flat to me (assuming light to propogate along geodesics in S_3) and then decrease and finally enclose me tightly; in this process all the air in the S_3-universe has passed through my mouth and nose once. An analogy is provided by a man "blowing up circles" on a sphere!

In a space of constant negative curvature, a nested set of 2-spheres constructed as above has no maximal member. On the contrary, the relation between area S and distance r from the origin is given by $S=4\pi a^2 \sinh^2(r/a)$, instead of $S=4\pi a^2 \sin^2(r/a)$ which holds for S_3. The areas and volumes of these 2-spheres thus increase faster than in the Euclidean case, and the total volume is of course infinite. Such spaces provide the constant-density space sections of open isotropic universes. On purely philosophical grounds most people probably hope that our universe turns out to be closed and thus finite. Infinite universes seem wasteful: for example, DNA patterns are finite and thus all of us would have infinitely many clones. But astronomical and other observations so far have not allowed us to determine the sign of the curvature.

SUGGESTED FURTHER READING

J. L. Synge, Relativity: The General Theory (North Holland, Amsterdam, 1960).

C. W. Misner, K. S. Thorne, J. A. Wheeler, Gravitation (W. H. Freeman and Co., San Francisco, 1973).

H. Stephani, General Relativity (Cambridge University Press, Cambridge, 1982).

ON THE RELATION BETWEEN GENERAL RELATIVITY AND QUANTUM THEORY: DIFFICULTIES AS SEEN BY A NON-EXPERT

(Summary of Talks by Jürgen Ehlers)

Given at a Workshop, Logan, February 86

The basic mathematical assumptions underlying classical general relativity and its physical interpretation are reviewed. Those structural features which arise from the non-existence of an a priori, given spacetime metric and associated isometry group are stressed, and it is emphasized that general relativity is not just a Poincare-invariant theory with a complicated gauge group. Then the fundamental assumptions on which special-relativistic quantum field theories have been based are considered and compared with general relativity from a mathematical and physical point of view. Two (old fashioned?) types of attempts to "quantize gravity" are then briefly discussed: (i) perturbation theory starting with linearized, free Einstein gravity, and (ii) canonical quantization starting from the ADM-dynamics. This is done only to illustrate difficulties which arise in "quantizing" gravity and to explain to quantum field theorists the point of view of at least one classically oriented relativist with the aim of improving communication and mutual understanding.

SOME IDEAS IN MODERN RELATIVITY PHYSICS

Arnold Rosenblum, Director

International Institute of Theoretical Physics

Logan, Utah

Relativity physics, including particles and fields, has come into prominence both with professional scientists and the educated public essentially within the last ten years. In the macroscopic regime, black holes, gravitational radiation, the binary pulsar and the early universe have been subjects of recent intense research activity. On the microscopic level, there has been an intense international effort to try to unify as a single superforce the fundamental interactions of nature. These include gravity and the unification occurs at extremely relativistic energies.

Older Results

I would like to begin this survey with macroscopic general relativity. Clearly from the form of Einstein's field equation

$$G^{\mu\nu} = T^{\mu\nu} \tag{1}$$

as a quasilinear hyperbolic field equation, we should expect gravitational radiation. From the fact that the equation is nonlinear, we should anticipate problems when we try to connect the radiation with material sources. In 1973 [1], Michael Reinhardt and I predicted the existence of a binary pulsar in a talk at the German Astrophysical Society. We pointed out that since approximately fifty percent of stars are in binary systems, it would not at all be surprising that a system consisting of two neutron stars should be found. A neutron star is a star of approximately the same mass as the sun but with a radius of only approximately thirteen kilometers. This is obviously a very compact object. There is the possibility of observing this compact system because associated with the neutron star is a magnetic field of approximately 10^{12} Gauss [2]. Because of electromagnetic radiation associated with the magnetic fields, it is possible to observe the period change of the binary system with a radio telescope. We pointed out that by observing this period change we could test indirectly for the existence of gravitational radiation in direct analogy to the classical electron spiraling in to the nucleus when giving off electromagnetic radiation. In 1975, the binary pulsar was found. In 1976 [3] Ehlers, Havas Goldberg and I pointed out that the Einstein quadrupole formula that was being used to compute the period change of the binary pulsar had not been established. Despite enormous amounts of work on the problem (see for example references [4], [5], [6] and [7]), it still has not been established in what cases the quadrupole formula holds and in particular, whether the compactness of the bodies involved effects the final result. Interestingly enough, some of the mathematical techniques developed for the problem of classical radiation reaction in general relativity have application in the problem of stochastic quantum field theory which will be discussed later.

There is still another indirect method to verify the existence of gravitational radiation. There are binary objects called cataclysmic binaries. Unlike the binary pulsar which is a clear system, there is an enormous amount of mass exchange. This mass exchange is governed by the Roche Lobe equipotential surface whose shape is a functional of the angular momentum of the system. For a cataclysmic binary the dominant effect driving off angular momentum is gravitational radiation. By using this mechanism it is possible to trace the evolution of this system and in particular its minimum period. For a recent reference see [8]. It should be pointed out that everything discussed so far is in connection with indirect observation of the effect of gravitational radiation and is not concerned with direct observation of the radiation. In the next section I will discuss more recent ideas for the detection of gravitational radiation and another effect called the "dragging of inertial frames" or the Lense Thirring effect.

Newer Methods

Up to this point neither the Lense-Thirring effect nor gravitational radiation has been observed directly. In this section I should like to propose how by using the "synchronization gap" in closed paths of clocks in accelerating coordinates systems or gravitational fields it is possible to test for these effects.

I would first like to start with a discussion of synchronization in special relativity. Assume that as in diagram 1 we have two clocks at rest at points a and b, respectively. If we wish to synchronize the clock at a with the clock at b using light signals, we first send from a a light signal to b' and then a light signal from b' to a'. We then send a message to clock b to set

$$t_{b'} = t_{a''} = \frac{t_a + t_{a'}}{2} \tag{2}$$

Still another way of synchronizing clocks at rest, the method of slow clocks transport, is shown in diagram 2.

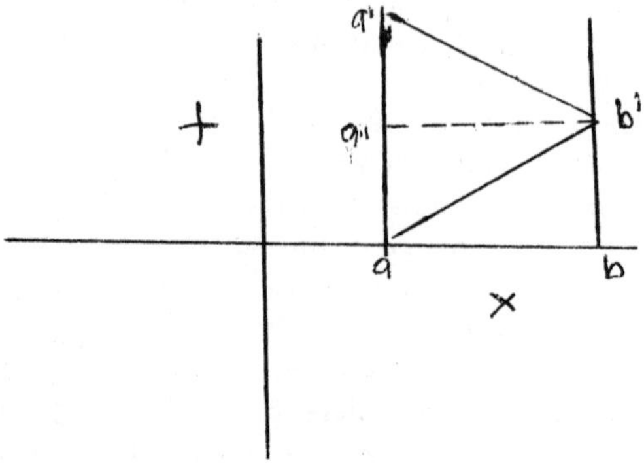

Fig. 1

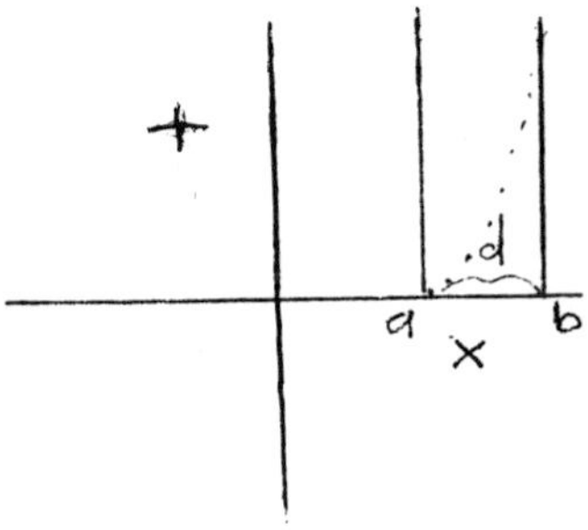

Fig. 2

We can simply take a third clock, compare its time with clock a and slowly transport it to clock b and then synchronize clock b. One might immediately ask is this slow clock transport synchronization equivalent to light signal synchronization for the case of two static clocks. I will show that this is the case. We have

$$d\tau^2 = dt^2 - \frac{dx^2}{c^2} - \frac{dy^2}{c^2} - \frac{dz^2}{c^2} \tag{3}$$

and then

$$d\tau = \sqrt{(1 - v^2/c^2)}dt \tag{4}$$

We have

$$d\tau - dt = (\sqrt{1 - v^2/c^2} - 1)dt \tag{5}$$

and $dt = d/v$ from diagram 2. Therefore we have

$$d\tau - dt = (\sqrt{1 - v^2/c^2} - 1)d/v \tag{6}$$

and in the low velocity limit we have

$$d\tau = dt \tag{7}$$

or

$$t = \tau = t_{b'} = t_{a''} \tag{8}$$

I would now like to discuss the problem of clock synchronization in the general case. We have

$$ds^2 = g_{\alpha\beta}dx^\alpha dx^\beta \tag{9}$$

or

$$ds^2 = g_{00}(dx^0)^2 + 2g_{oi}dx^0 dx^i + g_{ij}dx^i dx^j \tag{10}$$

If we simply complete the square in expression (1), we obtain

$$\begin{aligned} ds^2 &= g_{00}(dt + g_{00}^{-1} g_{0i}dx^i)^2 \\ &+ (g_{ij} - g_{00}^{-1} g_{0i}g_{0j})dx^i dx^j \end{aligned} \tag{11}$$

The second term is the well-known Landau-Lifshitz spatial line element associated with local light signal synchronization. The first term acts as the effective time or

$$'dt' = dt + g_{oo}^{-1} g_{oi}dx^i \tag{12}$$

We now integrate expression (12) along a closed path and obtain

$$'\Delta t' = \int_{\partial C} g_{oo}^{-1} g_{oi} dx^i \tag{13}$$

Let us apply the above formula to the case of clocks synchronized on a circle in a rotating frame. We have

$$\begin{aligned} ds^2 = & -\gamma^{-2}c^2(dt - r^2c^2\omega\gamma^2 \sin^2\theta d\phi)^2 + dr^2 + r^2 d\theta^2 + \\ & + r^2\gamma^2 \sin^2\theta d\phi^2 \end{aligned} \tag{14}$$

with $\gamma^{-2} = 1 - r^2\omega^2c^{-2}\sin^2\theta$

If we use expression (14) in formula (13) we obtain

$$'\Delta t' = 2\pi r^2 \omega c^{-2} \tag{15}$$

Using the radius of a geosynchronous orbit, r = 42,000 km and the ω of the earth $\omega = 7 \times 10^{-5}$ sec^{-1}, we obtain

$$'\Delta t' = 9\mu\text{sec} \tag{16}$$

which is an easily measurable quantity. Let us now examine the case of clock transport synchronization including gravitation. We have

$$ds^2 = -A^2dt^2 + B^2dr^2 + r^2d\theta^2 + r^2\sin^2\theta d\phi^2 \tag{17}$$

with $A^2 = B^{-2} = 1 - \frac{2m}{r}$

We rewrite the above as

$$d\tau^2 = dt^2 \left\{ A^2 - B^2 \left[\left(\frac{dr}{dt}\right)^2 - r^2 \left(\frac{d\theta}{dt}\right)^2 - r^2\sin^2\theta \left(\frac{d\phi}{dt}\right)^2 \right] \right\} \tag{18}$$

Taking the weak-field, slow-motion limit of the above, we obtain

$$d\tau = dt = \left[\phi - v^2/2\right] dt \tag{19}$$

and integrating the above along a closed path, we obtain

$$\Delta(t - \tau) = c^{-2} \int_{t_a}^{t_b} (1/2v^2 - \phi) dt \tag{20}$$

with $v/c \ll 1$ and $\phi c^{-2} = -Gm/rc^2 \ll 1$. If we apply the above result for one revolution about a circular orbit, we obtain (9)

$$\Delta(t - \tau) = 3\pi r^2 \omega c^{-2} \tag{21}$$

This synchronization gap clearly differs from that obtained by using light signal synchronization. If we use clock transport synchronization around an ellipse we obtain

$$\Delta(t - \tau) = 3\pi/c^2 (GMa)^{\frac{1}{2}} \tag{22}$$

where a is the semi-major axis. For the case of light signal synchronization, we would clearly like to relax the restriction that the clocks must be infinitesimally close to one another. We notice that for clocks on the vertices of an N-sided regular polygon rotating with angular velocity ω we obtain (10)

$$\Delta = r^2\omega c^2 \, N \sin(2\pi/N) \tag{23}$$

which of course reduces in the large N limit to the case of the circle.

To discuss the case of using light signal synchronization to detect dragging of inertial frames, we start from the Cohen-Brill metric

$$-d\tau^2 = A^2dt^2 + B^2dr^2 + r^2(d\phi - \Omega dt)^2 \tag{24}$$

where $A^2 = B^{-2} = 1 - 2m/r$, $\Omega = 2J/R^3 - \omega_o$ and $J = KM_* R_*^2 \omega$ and $K \approx 1$ with ω_o the rotation rate of an inertial frame as measured by an observer at infinity and with J the body's angular momentum. The synchronization gap expressed in terms of proper time for orbiting satellites in circular equatorial orbits is

$$\Delta S = 2\pi r^2 \omega(1 + 3/2M/r - (2kM)/r(R_*/r)^2) \tag{25}$$

where the first term in the parenthesis is the special relativistic one, the second term arises from a term similar to the gravitational red shift, the third term from the dragging of inertial frames term.

For experimental purposes, it is clearly better to eliminate the lower order terms in expression (25). To do this we send up clocks in satellites both in geosynchronous and anti-geosynchronous orbits. We synchronize using light signals, first the clocks in the geosynchronous orbits and then by using light signals going the other way. The simple addition of the synchronization gap of both the geosynchronous and anti- geosynchronous orbits will isolate the effects of dragging of inertial frames and with possible absolute clock accuracies of 1 part in 10^{18} in the next five years [11] leads to the possibility of experimental determination of dragging of inertial frames. In geosynchronous orbit, the dragging of inertial frames for 1 revolution leads to a synchronization gap of approximately 1.92×10^{-17} seconds for $r = R_*$.

To summarize, I have presented a new way using the latest in clock technology to measure dragging of inertial frames. It is also clear that we can use the above method to test for gravitational radiation and the fifth force. This will, of course, require even greater accuracy. It should be very strongly pointed out that the use of light signals is not necessary when using the lack of synchronization of clocks because the method of clock transport synchronization can be used which required, in principle, no light signals.

Stochastic Quantum Field Theory

On the face of it, there would seem to be no connection between classical equations of motion and the approximation methods used to solve them and quantum field theory and the approximation methods used to solve the quantum equations of motion. Neither canonical quantization nor path integral quantization make use of classical equations of motion. The recently developed method of stochastic quantum fields by Parisi and Wu [12] makes use of classical-like equations of motion before calculating the quantum propagator. It is therefore possible to make use of techniques developed to solve classical nonlinear equations of motion which are not merely a simple linearization in problems of quantum field theory. In particular, one of the aims of the approach is to carefully examine the problem of non-renormalizability of quantum gravity and Hawking radiation.

The expectation value of a functional $f[\phi]$ of the Euclidean quantum field ϕ is given by the path integral

$$\langle f[\phi]\rangle = \int D[\phi]\, f[\phi]\, P[\phi] \tag{26}$$

with

$$P[\phi] = e^{-1/h S_E[\phi]} \Big/ \int D[\phi]\, e^{-1/h S_E[\phi]}$$

The field ϕ may be viewed as an equilibrium process characterized by the Euclidean action $S[\phi]$ and the effective temperature h/K. This is possible because $P[\phi]$ is a positive probability distribution. Parisi and Wu constructed an effective relaxation process $\phi[x,s]$ such that $\phi(x)$ is obtained as $s \rightarrow \infty$. We have

$$\frac{\partial\phi(x,s)}{\partial s} = \frac{-\delta S_E[\phi]}{\delta\phi(x,s)} + h^{1/2}\xi(x,s) \tag{27}$$

with

$$\langle \xi(x,s)\xi(x',s')\rangle = 2\delta(s-s')\delta^4(x-x') \tag{28}$$

In addition, the Euclidean quantum field theoretical expectation values are given by

$$\langle f[\phi(x)]\rangle_{QFT} = \lim_{s\to\infty} \langle f[\phi(x,s)]\rangle \tag{29}$$

Equation (27) is a classical field equation of the sort encountered in solving problems in classical non-linear [13-15] equations of motion. Work is now under way to apply this method to the two dimensional Kink solution in the 2-dimensional ϕ^4 theory [16]. The program is use the wisdom obtained from solving this problem to that of quantum gravity and Hawking radiation.

Conclusion

Relativity physics is now in excellent health. On the macroscopic level, improvement in clock accuracies in the near future promise more tests of classical general relativity. On the microscopic level, new approaches such as a Parisi-Wu stochastic quantum field theory combined with modern methods of applied mathematics such as matched asymptotic expanisons lead to the hope of some progress on the outstanding problems associated with quantum gravity and supergravity.

References

1. The General Relativistic Two-Body Problem, A. Rosenblum, M. Reinhardt, Mitt. Astron. Ges., No 34, 19 (1973).
2. Self-Consistent Pulsar Magnetospheres in General Relativity, A. Rosenblum, J. Cohen, Ap. J., 186, (1973).
3. Comments of Radiation Damping and Energy Loss in Binary Systems, J. Ehlers, A. Rosenblum, J.N. Goldberg, P.H. Grap, Ap. J. Lett., 201, L77 (1970).
4. The Gravitational Energy Loss in Scattering and the Einstein Quadrupole Formula, A. Rosenblum, Phys. Rev. Lett., Oct 9, (1972).
5. Gravitational Radiation Energy Loss in Scattering Problems, A. Rosenblum, Phys. Lett. A., 81A, 1 (1981).

6. The Third Order Equations of Motion in the Mast Motion Approach in General Relativity, A. Rosenblum, Phys. Lett. A., 93A, II (1983).
7. Energy Loss Due to Small Angle Scattering in General Relativity, J. Phys. A. Math. Gen. 16, 2751 (1983), Rosenblum.
8. Evolution of Rotationally and Tidally Distorted Low-Mass Close Binary Systems, L.A. Nelson, W.Y. Chau and A. Rosenblum, Ap. J. 299, 658-667, (1985).
9. Clock-Transport Synchronization in Non-inertial Frames and Gravitational - Fields, J. Cohen, H. Moses, A. Rosenblum, Phys. Rev. Lett. (1983).
10. Electromagnetic Synchronization of Clocks with Finite Separation in a Rotating System, J. Cohen, H. Moses, A. Rosenblum, Class. Quan. Grav. 1 (1984) L57-59.
11. H. Dehmelt, Ann. Phys. Fr. 10, (1985) 777-795.
12. G. Parisi, Y.S. Wu, Sci. Sinica 24 483 (1981).
13. Radiation Damping of Color in Classical Yang-Mills Theory, R. Kates, A. Rosenblum, Phys. Rev. 28, 3066 (1983).
14. Theory of Motion for Monopoles, Dipole Singularities of Classical Yang-Mills-Higgs Field A: Laws of Motion. W. Drechsler, P. Havas, A. Rosenblum, Phys. Rev., Feb 1984.
15. Theory of Motion for Monopole-Dipole Singularities of Classical Yang-Mills-Higgs Fields B: Approximation Scheme, Equations of Motion, W. Drechsler, P. Havas, A. Rosenblum, Phys. Rev., Feb 1984.
16. R. Kates, A. Rosenblum (in preparation).

COSMOLOGICAL VACUUM CONFIGURATION IN SUPERSTRING THEORIES WITH ONLY THREE-SPACE EXPANDING

Zi Wang and V. Gordon Lind
Department of Physics
International Institute of Theoretical Physics
Utah State University
Logan, UT 84322, USA

Yong-Shi Wu
Department of Physics
University of Utah
Salt Lake City, UT 84112, USA

ABSTRACT

Cosmological compactification of superstring theories is studied in the field theory limit, with the quadratic Euler-Gauss-Bonnet term included in the gravitational action. We find a vacuum configuration, which is of the Robertson-Walker parameter $k = -1$, of three-space scale factor $R_3(t) \sim t$ and of constant internal-space scale factor ($R_6(t) = $ const.); otherwise, it is the same as the static Candelas-Horowitz-Stominger-Witten (CHSW) configuration. While the CHSW configuration is unstable, our three-space expanding configuration is stable against time-dependent purturbations in $R_3(t)$ and $R_6(t)$. These results indicate the existence of cosmological models (with the presence of matter) in superstring theories which predict, asymptotically in the limit of infinitely dilute matter, no time variation for the coupling constants in four dimensions.

In recent years, it has become clear that superstring theories [1,2], especially the $E_8 \times E_8$ heterotic superstring [3], are promising candidates of quantum gravity and could be a "theory of everything," i.e., a consistent quantum theory unifying all known interactions. The low energy field theory limit of superstring theories is a modified version of the ten-dimensional supergravity coupled to supersymmetric Yang-Mills fields [4]. Some of the modifications are due to the fact that the tree-level gravity in string theories is different from the Einstein gravity. In the field theory limit, this leads,in addition to the Einstein action, to terms which are quadratic [3, 5] and higher [6] in the curvature. The ghost-free feature of superstring theories suggests some particular form for these additional terms in the gravitational action [7]. Especially, the leading quadratic terms should be of the Euler-Gauss-Bonnet form [5, 7]. Also the ten-dimensional supersymmetry relates the presence of the Euler-Gauss-Bonnet term to that of the Lorentz-Chern-Simons term which was added by Green and Schwarz [2] to the field theory for the purpose of anomaly cancellation and can be deduced from string theories [3, 8]. In fact, one can derive one of the two terms from the presence of the other by supersymmetrizing the field theory action [9, 10]. Thus, in string theories there arises naturally the problem of what are the physical consequences of the additional terms, e.g., Euler-Gauss-Bonnet term, in gravitational phenomena. In ref. [11] some striking properties have been found in the string-generated gravity model, especially for the Schwarzschild-like solutions which may be of relevance to the behavior of black holes near its would-be singularity. In what follows we study the cosmological compactification [12] in the string theory context, with the Euler-Gauss-Bonnet term included. Before doing this let us provide more motivation for doing so in the next paragraph.

Obviously for a ten-dimensional theory to be realistic it is necessary that the vacuum background be of the form $M_4 \times K_6$, where M_4 is the four-dimensional space-time and K_6 is some compact six-dimensional manifold. Candelas, Horowitz, Strominger, and Witten (CHSW) [12] have found a candidate for vacuum configuration in superstring theories that has an unbroken N-1 supersymmetry in four dimensions and a vanishing cosmological constant. Their compactification is a static one with M_4 being Minkowski space-time and K_6 a Calabi-Yau manifold [13]. An alternative, which is of interest in cosmology, is the so-called cosmological compactification, namely time-dependent vacuum configuration with six dimensions compactified. In an expanding universe like ours, when the density of matter becomes so dilute due to enough expansion that its effects can be neglected for local physics, it is quite likely that the vacuum background may still be a time-dependent one. The idea that the vacuum background in cosmology might well be time-dependent is not new. In the early days of cosmology the de Sitter model suggested an empty but expanding universe [14]. However, in the recently popular Kaluza-Klein-like cosmology there is an old, well-known problem, namely, how to keep the size of the internal space fixed while having the ordinary three-space expanding. This is because on one hand, according to Kaluza-Klein reduction, the coupling constants in four dimensions depend on the size of the internal space [15, 16], and, on the other hand, there has been no observational evidence for time variation of these coupling constants [17]. Up to now no good solution to this problem has appeared in the literature about Kaluza-Klein cosmology. It is worth re-examining the same problem in string cosmology since the bosonic sector in string theories which is relevant to the cosmological compactification contains several fields other than the metric field, and there may be a chance for a conspiracy among these bosonic fields against time variation of the internal space.

Indeed, we have found a space-expanding vacuum configuration in the field theory limit of superstring theories which is of Robertson-Walker parameter $k = -1$, of three-space scale factor $R_3(t)$ proportional to the age of the universe and of constant six internal-space scale factor ($R_6(t)$ = const.); otherwise it is the same as the static CHSW configuration. We have also studied the stability for both our time-dependent and the static CHSW configuration. It turns out that while the latter is unstable the former is stable against time-dependent perturbations in $R_3(t)$ and $R_6(t)$. Since a cosmological vacuum configuration can be viewed as a limiting situation of some cosmological models with the presence of matter when matter becomes infinitely dilute due to enough expansion of the universe, our results indicate the existence of cosmological models with the presence of matter in superstring theories which predict constant coupling constants in four dimensions at least asymptotically.

The bosonic sector of the action in the field theory limit of superstring theories with the Euler-Gauss-Bonnet term included is given by

$$S = \int d^{10} x \, -g \left\{ R - 3/2\phi^{-3/2} H_{MNP}^{2} - 9/8(\phi^{-1}\partial_m\phi)^2 \right.$$

$$\left. - 1/2\phi^{-3/3} [1/30 TrF_{MN}^{2} - (R_{MNPQ}^{2} - 4R_{MN}^{2} + R^2)] \right\} \quad (1)$$

with signature (-, +, ..., +), where ϕ, F_{MN} and H_{MNP} are the dilation, Yang-Mills field strength and the field-strength associated with the Kalb-Rammond potential. Varying g_{AB}, B_{MM}, A_M^a and ϕ we obtain the ten-dimensional equations of motion:

$$R_{AB} - 1/2 g_{AB} R = 9/2\phi^{-3/2}[H_{AMN} H_B^{MN} - 1/6 g_{AB} H^2_{MNP}]$$

$$+ 9/8\phi^{-2}[\partial_A\phi\partial_B\phi - 1/2 g_{AB}(\partial_M\phi)^2] + 1/30\phi^{-3/4}[TrF_{AM}F_B^{\ M} - 1/4 g_{AB} TrF_{MN}^{\ 2}]$$

$$+1/2\phi^{-3/4}[1/2 g_{AB}(R^2_{MNPQ} - 4R_{MN}^{\ 2} + R^2) - 2RR_{AB} + 4R_{AM}R_B^{\ M} + 4R_{AMBN}R^{MN}$$

$$- 2R_A^{\ MNP}R_{BMNP}] + 9\nabla^M(\phi^{-3/2}H_{APQ}R_{MB}^{\ \ PQ}) \quad (2)$$

$$\nabla_M(\phi^{-3/2}H^{MNP}) = 0 \quad (3)$$

$$D_M(\phi^{-3/4}F^{MPa}) + 9(\phi^{-3/2}F_{MN}{}^{a}{}_{H}{}^{MNP}) = 0 \quad (4)$$

$$6\phi^{-5/2}H^2_{MNP} + 6\phi^{-3}(\partial_M\phi)^2 + 6\nabla_M(\phi^{-2}\partial^M\phi)$$

$$+ \phi^{-7/4}[1/3P TrF_{MN}^{\ 2} - (R^2_{MNPQ} - 4R_{MN}^{\ 2} + R^2)] = 0 \quad (5)$$

where ∇_M is the ordinary Levi-Civita covariant derivative and D_M the doubly covariant derivative including the Yang-Mills potential. For any compactification, the following Bianchi identity must be fulfilled [18]

$$dH = trR\wedge R - 1/30 TrF\wedge F \quad (6)$$

To achieve a cosmological compactification, we assume the metric in ten-dimensions to be of the generalized Robertson-Walker form

$$g_{MN} = \begin{pmatrix} -1 & & \\ & R_e^{\ 2}(t)\tilde{g}_{ij}(x) & \\ & & R_6^{\ 2}(t)\tilde{g}_{mn}(y) \end{pmatrix} \quad (7)$$

where M.N = 0, 1,...9; i, j = 1, 2, 3; m,n = 4, 5,...9;

$$g_{ij}(t.x) = R_3^{\ 2}(t)\tilde{g}_{ij}(x); \; g_{mn}(t,y) = R_6^{\ 2}(t)\tilde{g}_{mn}(y).$$

$R_3(t)$, $R_6(t)$ are the scale factors of the ordinary 3-space and of the internal 6-space respectively. $\tilde{g}_{ij}(x)$ is maximally symmetric in 3-space and $\tilde{g}_{mn}(y)$ is a Calabi-Yau metric.

In action (1) the gauge coupling constant g_{10} has been absorbed by rescaling ϕ [18, 19] (gravitational coupling constant $K_{10} = 1$). In fact, (or some power of it) acts as a coupling constant because a certain power of ϕ appears in front of each term in eq. (1) except the Einstein term. The observational limits on the variation of the coupling constants with time suggest that one can take ϕ to be constant as a good approximation. In the static case, supersymmetry in four dimension implies that Hmnp are vanishing and Yang-Mills fields only survive in the internal space [12]. In the time-dependent case it would be possible for some spatial components of H to be non-zero because the metric (7) breaks the maximal symmetry of four-dimensional space-time. However, in order to avoid an unwanted cosmological constant, we also assume that all components of H are vanishing. Moreover, we require that Yang Mills fields still only survive in the internal space in order to preserve the gauge invariance in four dimensions.

To summarize, we take the following simplest ansatz

$$H_{MNP} = o \tag{8a}$$

$$F_{MN} = F_{mn}, \; M = m, \; N = n \tag{8b}$$

$$= o, \text{ otherwise}$$

$$\phi = \text{const.} \tag{8c}$$

Using equation (8a), the Bianchi identity (6) becomes

$$1/30 \; \mathrm{Tr} F \wedge F = \mathrm{tr} R \wedge R \tag{9}$$

and the Yang-Mills field equation (4) reduces to

$$D_M(\phi^{-3/4} F^{MPa}) = 0 \tag{10}$$

Moreover, the Kalb-Ramond field equation (3) is satisfied automatically.

In the metric (7) the metric for the internal space is described by $g_{mn}(t,y) = R_6^2(t)\, \tilde{g}_{mn}(y)$, where $\tilde{g}_{mn}(y)$ is a Ricci-flat Kahler metric, i.e., a Calabi-Yau metric. Such a metric may be called conformally Calabi-Yau. Using the conformal transformation [20] one can easily show that both sides of (9) are conformally invariant and a solution of Yang-Mills field equation (10) on a Calabi-Yau space in the static case is also a solution for the time-dependent case. The ansatz (8a) and (8c) make the dilation field equation (5) a constraint equation

$$1/30 \; \mathrm{Tr} F_{MN}^2 = R_{MNPQ}^2 - 4R_{MN}^2 + R^2 \tag{11}$$

and by using eq. (11) and (8) one can show that the ten dimensional scalar curvature is zero ($R = 0$).

Thus, under the ansatz (8) the generalized Einstein equations (2) are reduced to

$$R_{AB} = \phi^{-3/4}\Big[1/30 \mathrm{Tr} F_{AM} F_B{}^M + 2R_{AM} R_B{}^M + 2\, R_{AMBN} R^{MN} - R_A{}^{MNP} R_{BMNP}\Big] \tag{12}$$

In terms of $R_3(t)$ and $R_6(t)$ equations (12) can be further written as

$$\frac{\ddot{R}_3}{R_3} + 2\,\frac{\ddot{R}_6}{R_6} = \phi^{-3/4}\left[4\frac{k\ddot{R}_3}{R_3^{\,3}} + 6\frac{\ddot{R}_3^{\,2}}{R_3^{\,2}} + 24\frac{\ddot{R}_6^{\,2}}{R_6^{\,2}} + 24\frac{\ddot{R}_3\ddot{R}_6}{R_3R_6} + 4\,\frac{\ddot{R}_3\dot{R}_3^{\,2}}{R_3^{\,3}} + 20\,\frac{\ddot{R}_6\dot{R}_6^{\,2}}{R_6^{\,3}} + 12\,\frac{R_3R_6}{R_3R_6}\left(\frac{R_3}{R_3} + \frac{R_6}{R_6}\right)\right] \tag{13a}$$

$$\frac{2k}{R_3^{\,2}} + \frac{\ddot{R}_3}{R_3} + 2\,\frac{\dot{R}_3^{\,2}}{R_3^{\,2}} + 6\,\frac{\dot{R}_3\dot{R}_6}{R_3R_6} = \phi^{-3/4}\left[12\,\frac{k^2}{R_3^{\,4}} + 12\,\frac{k\ddot{R}_3}{R_3^{\,3}} + 24\,\frac{k\dot{R}_3^{\,2}}{R_3^{\,4}} + 72\,\frac{k\dot{R}_3\dot{R}_6}{R_3^{\,3}R_6} + 6\,\frac{\ddot{R}_3^{\,2}}{R_3^{\,2}} + 12\,\frac{\ddot{R}_3\ddot{R}_6}{R_3R_6} + 12\,\frac{\ddot{R}_3\dot{R}_3^{\,2}}{R_3^{\,3}} + 24\frac{\ddot{R}_3\dot{R}_3\dot{R}_6}{R_3^{\,2}R_6} + 12\,\frac{\ddot{R}_6\dot{R}_3\dot{R}_6}{R_3R_6} + 12\,\frac{\dot{R}_3^{\,4}}{R_3^{\,4}} + 72\,\frac{\dot{R}_3^{\,3}\dot{R}_6}{R_3^{\,3}R_6} + 96\,\frac{\dot{R}_3^{\,2}\dot{R}_6^{\,2}}{R_3^{\,2}R_6^{\,2}} + 60\,\frac{\dot{R}_3\dot{R}_6^{\,3}}{R_3R_6^{\,3}}\right] \tag{13b}$$

$$\frac{\ddot{R}_6}{R_6} + 5\,\frac{\dot{R}_6^{\,2}}{R_6^{\,2}} + 3\,\frac{\dot{R}_3\dot{R}_6}{R_3R_6} = -\phi^{-3/4}\left[6\,\frac{k^2}{R_3^{\,4}} + 8\,\frac{k\ddot{R}_3}{R_3^{\,3}} + 12\,\frac{k\dot{R}_3^{\,2}}{R_3^{\,4}} + 36\,\frac{k\dot{R}_3\dot{R}_6}{R_3^{\,3}R_6} + 6\,\frac{\ddot{R}_3^{\,2}}{R_3^{\,2}} + 18\,\frac{\ddot{R}_3\ddot{R}_6}{R_3R_6} + 12\,\frac{\ddot{R}_6^{\,2}}{R_6^{\,2}} + 8\,\frac{\ddot{R}_3\dot{R}_3^{\,2}}{R_3^{\,3}} + 18\,\frac{\ddot{R}_3\dot{R}_3\dot{R}_6}{R_3^{\,2}R_6} + 12\,\frac{\ddot{R}_6\dot{R}_3\dot{R}_6}{R_3R_6^{\,2}} + 10\frac{\ddot{R}_6\dot{R}_6^{\,2}}{R_6^{\,3}} + 6\,\frac{\dot{R}_3^{\,4}}{R_3^{\,4}} + 36\,\frac{\dot{R}_3^{\,3}\dot{R}_6}{R_3^{\,3}R_6} + 48\,\frac{\dot{R}_3^{\,2}\dot{R}_6^{\,2}}{R_3^{\,2}R_6^{\,2}} + 30\,\frac{\dot{R}_3\dot{R}_6^{\,3}}{R_3R_6^{\,3}}\right] \tag{13c}$$

In the static case, (9) is satisfied [12] by setting the gauge field on the manifold K_6 equal to the spin connection for K_6 with a suitable embedding of SU(3) holonomy group in SO(32) or $E_8 \times E_8$.

$$F_{mn}{}^{\alpha\beta} = R_{mn}{}^{\alpha\beta} \tag{14}$$

where idices m, n,... are curved indiced for K_6, α, β,... are flat indices. In the time-dependent case the spin connection is invariant under a conformal transformation $g_{mn}(y) \quad g_{mn}(t,y) = R^2_6(t)\ g_{mn}(y)$ so far as its components entirely on the compact maniford K_6 are concerned. Therefore, eq. (14) is also a solution to the constraint (9) for the time-dependent case.

Putting eq. (14) into the constraint equation (11) one obtains

$$3\frac{k^2}{R_3^4} + 4\frac{k\ddot{R}_3}{R_3^3} + 6\frac{k\dot{R}_3^2}{R_3^4} + 24\frac{k\dot{R}_3\dot{R}_6}{R_3^3R_6} + 3\frac{\ddot{R}_3^2}{R_3^2} + 12\frac{\ddot{R}_3\ddot{R}_6}{R_3R_6} + 12\frac{\ddot{R}_6^2}{R_6^2} - 4\frac{\ddot{R}_3\dot{R}_3^2}{R_3^3}$$
$$+ 12\frac{\ddot{R}_3\dot{R}_3\dot{R}_6}{R_3^2R_6} + 12\frac{\ddot{R}_6\dot{R}_3\dot{R}_6}{R_3R_6^2} + 20\frac{\ddot{R}_6\dot{R}_6^2}{R_6^3} + 3\frac{\dot{R}_3^4}{R_3^4} + 24\frac{\dot{R}_3^3\dot{R}_6}{R_3^3R_6} + 54\frac{\dot{R}_3^2\dot{R}_6^2}{R_3^2R_6^2}$$
$$+ 60\frac{\dot{R}_3\dot{R}_6^3}{R_3R_6^3} + 45\frac{\dot{R}_6^4}{R_6^4} = 0 \tag{15}$$

Now we are in a position to seek possible solutions of Eqs. (13) and (15). For the power-law solutions write

$$R_3(t) = \alpha_3 t^{\beta_3}$$
$$R_6(t) = \alpha_6 t^{\beta_6} \tag{16}$$

where the constants α_3, α_6, β_3, and β_6 are to be determined.

Because of the terms in k in eqs. (13) and (15), where k = +1, 0, -1, we do not have much choice if we want to satisfy all the equations. One can see that the only possible choices are β_3, β_6 = 1 or 0. Then the power-law solutions are only the following two:

$$\text{(i)} \begin{cases} R_3(t)=t + t_o \\ R_6(t)=\text{const.} \end{cases} \quad \text{for } k = -1 \tag{17a}$$

$$\text{(ii)} \begin{cases} R_3(t)=\text{const.} \\ R_6(t)=\text{const.} \end{cases} \quad \text{for } k = 0 \tag{17b}$$

where t_o is the integration constant. The solution (ii) is just the static CHSW configuration [12]. The solution (i) represents a cosmological vacuum in the late universe. Quite likely these are the only solutions, but we have not been able to prove it.

To study stability of the solutions we first consider time-dependent perturbation in $R_3(t)$ and $R_6(t)$. For the solution (17a) we write

$$
\begin{aligned}
R_3(t) &= t + \gamma_3(t) \\
R_6(t) &= R_{60} + \gamma_6(t)
\end{aligned}
\tag{18}
$$

Substituting (18) into eqs. (13) we obtain, up to first-order in γ_3 and γ_6,

$$
\begin{aligned}
\ddot{\gamma}_3 + 2t^{-1}\gamma_3 &= o \\
\ddot{\gamma}_6 + 3t^{-1}\gamma_6 &= o
\end{aligned}
\tag{19}
$$

with the solution

$$
\begin{aligned}
\gamma_3(t) &= \gamma_{30} - \gamma_{30} t^{-1} \\
\gamma_6(t) &= \gamma_{60} - \gamma_{60} t^{-2}
\end{aligned}
\tag{20}
$$

where γ_{30}, γ_{30}, γ_{60}, and γ_{60} are all constants. Therefore, the solution (17a) is critically stable against time-dependent perturbations in $R_3(t)$ and $R_6(t)$.

Similarly, for the solution (17b) we obtain the equations for perturbations as

$$
\begin{cases}
\ddot{\gamma}_3 = o \\
\ddot{\gamma}_6 = o
\end{cases}
\tag{21}
$$

It is clear that the static CHSW configuration (17b) is critically unstable.

To conclude, some remarks about our cosmological vacuum configuration (17a) are in order. First of all, for this configuration the action (1) vanishes so it has a zero cosmological constant. Secondly, this configuration differs from the static CHSW one only by the feature that three-space is expanding. Therefore, it is expected to lead to the same physics as the CHSW vacuum for phenomena local in four dimensional space-time, i.e., those occuring in regions which are small compared to both cosmological length and time scales. Thirdly, whether our three-space expanding vacuum leads to a supersymmetry breaking on the cosmological time scale remains to be seen. The study of this problem is beyond the present paper and worthy of a separate publication. Fourthly, we observe that in (17a) the time dependence of the scale factor in three-space, namely $R_3(t) \sim t$, coincides with the large-t behavior of $R_3(t)$ in the standard Robertson-Walker-Friedman cosmology with $k = -1$ (see eq. reference [14]). This result is nontrivial and interesting since in string cosmology we have extra dimensions to be compactified and more bosonic fields than in standard cosmology. Finally, and most importantly, our time-dependent vacuum (17a) has the following advantages in the context of cosmology, especially of Kaluza-Klein-like cosmology: it both incorporates the observed (three-space) expansion of the universe and predicts the time non-variation of the coupling constants in four dimensions. It is very likely that one can construct superstring-generated cosmological models with the presence of matter which, asymptotically in the infinitely-dilute limit of matter, reproduce our cosmological vacuum configuration or other ones with only three-space expanding. In this way, superstring theories seem to be promising to solve an old problem in Kaluza-Klein cosmology. The study of cosmological models with the presence of matter in superstring theories is ongoing in this regard. We notice that in ref. [21,22] the same physical problem was also considered, but not in the context of superstring theories.

ACKNOWLEDGEMENTS

Z. Wang and V. G. Lind would like to thank I. Anderson, S. Raby, A. Rosenblum, and R. C. Slansky for useful discussions. Y. S. Wu thanks V. G. Lind of Utah State University for his warm hospitality. This work was supported in part by NSF grant PHY-8405648.

REFERENCES

1. J. H. Schwarz, Phys. Rep. 89 (1982), 223; M. B. Green, Surv. High Energy Phys. 3 (1983), 127.
2. M. B. Green and J. H. Schwarz, Phys. Lett. 149B (1984), 117.
3. D. J. Gross, J. A. Harvey, E. Martinec, and R. Rohm, Phys. Rev. Lett. 54 (1984), 502; Nucl. Phys. B256 (1985), 253.
4. A. H. Chamseddine, Nucl. Phys. B185 (1981), 403; E. Bergshoeff, M. deRoo, B. deWit, and P. Van Nieuwenhuizen, Nucl. Phys. B195 (1982), 97; G. F. Chapline and N. S. Manton, Phys. Lett. 120B (1983), 105.
5. B. Zwiebach, Phys. Lett. 156B (1985), 315.
6. D. J. Gross and E. Witten, Princeton preprint (1986).
7. B. Zumino, UCB-PTH-85/13, LBL-19302 preprint (1985).
8. C. Hull and E. Witten, Phys. Lett. 160B (1985), 398; A. Sen., Phys. Lett. 166B (1986), 300.
9. L. Romans and N. Warner, CALT preprint 68-1291 (1985). C. Cecotti, S. Ferrara, L. Girardello, and M. Porrati, CERN preprint, TH4253/85 (1985).
10. D. Deser, in Proceedings of the Workshop on Supersymmetry and its Applications, Cambridge 1985 (Cambridge University Press, Cambridge, Mass.).
11. D. G. Boulware and S. Deser, Phys. Rev. Lett. 55 (1985), 2656.
12. P. Candelas, G. T. Horowitz, A. Strominger and E. Witten, NSF-ITP preprint 170 (1984), to appear in Nucl. Phys. B.
13. E. Calabi, in Algebraic Geometry and Topology, A Symposium in Honor of S. Lefschetz (Princeton University Press, 1957) p. 78; S. T. Yau, Proc. Natl. Acad. Sci., 74 (1977), 1978.
14. S. Weinberg, Gravitation and Cosmology, part five, John Wiley & Sons, 1972.
15. Y. S. Wu, Acta Physica Sinica, 29, (1980) 395.
16. S. Weinberg, Phys. Lett., 125B (1983), 265.
17. See e.g. W. J. Marciano, Phys. Rev. Lett., 52 (1984), 489; W. McCrea, M. J. Rees and S. Weinberg, eds., The constants of Nature, Phil. Trans., to be published.
18. E. Witten, Phys. Lett. 149B (1984), 351.
19. M. Dine and N. Seiberg, Phys. Rev. Lett. 55 (1985), 366.
20. C. N. Yang, Phys. Rev., D16 (1977), 330.
21. P. G. O. Freund and P. Oh, Nucl. Phys. B255 (1985), 688.
22. G. F. Chapline and G. W. Gibbons, Phys. Lett. 135 (1984), 43.

A POWERFUL NEW PROGRAMMING MODEL FOR PARALLEL COMPUTATION†

Andrew M. Kobos, Randy E. VanKooten and Martin A. Walker

Myrias Research Corporation
Edmonton, Alberta, Canada

Abstract

If a looping construct has the property that no iteration of the loop requires for its execution a result from a previous iteration, correct results can be obtained by executing all of the loop iterations simultaneously on different processors.

At Myrias Research Corporation, a virtual machine has been designed, and is being implemented, on which such parallel execution of loop constructs is achieved by replacing the looping instruction ("**do**" in ANSI Fortran 77) with a modified instruction ("**par do**" in Myrias Parallel Fortran – MPF). Each loop iteration executes in its own, separate memory space, and these memory spaces are automatically merged when all of the tasks have completed. The architecture of the Myrias parallel computer is described briefly.

We describe a new memory model which utilizes local, distributed memory, and allows for the dynamic reconfiguration of parallel tasks at the operating system level. Such a model gives rise to a powerful new parallel programming method. Further, we describe how recursive parallel methods (RPM) can be used effectively. The three main language extensions are orthogonal, and their combination provides easy access to flexible, high order, adaptive algorithms. Algorithmic examples of the application of this programming model are demonstrated.

1. Introduction

As vector processing computers begin to approach the limits of computational speed imposed by technology and physics, major computer manufacturers, and the scientific computing community in general, have begun to realize that further, substantial gains in processing speed (factors of greater than 10) could only be achieved by linking a number of processors together, so that the sequential power of individual processors could be harnessed into a single, parallel multiprocessor computer. Each processor could work in parallel on independent tasks, contributing to the fast solution of large problems.

One, now operational, solution is to link several high speed, pipelined vector processors. In this case, however, hardware restrictions have been propagated to the

† This paper has been left virtually unchanged from its original version written in 1985. Since then, Myrias Research Corporation has developed a working prototype encompassing many of the features discussed herein, and is presently completing an advanced, commercially available, parallel system.

user programming model. As the result of pipelining being employed in vector processors, their performance is sensitive to the form of algorithms involved in the problem to be solved. Not every computationally intensive application problem encountered in realistic physical modeling and simulation, in real time signal processing, etc., is amenable to vectorization. A much greater number of such problems are quite susceptible to parallel processing. In general, vectorizable computational problems constitute only a subset of those which can be solved in parallel.

Another approach to parallel computing, now under development in a number of establishments, is to connect in some way a large number of slower (and cheaper) processors so that a large number of computational tasks could be executed concurrently, and thus considerable performance gains be achieved. Such a notion stems from the fact that in a large number of important applications, independent computational tasks can be easily discerned [1]. This approach, however, can create inter-task communication problems.

Myrias Research Corporation has taken the latter approach to multiprocessing 'to the limit', and is engaged in the construction of a parallel computing system [2] that will allow a large number of heterogeneous tasks to be computed simultaneously. No restrictions are imposed on either the task's nature or computational expression.

2. General description of the Myrias computer system

2.1 Architecture of the Myrias Parallel Computer

The parallel computer under development at Myrias Research Corporation has a multilayered architecture. The layers are labeled as follows: hardware, control firmware, operating software, and programming languages. Each of these will be described briefly below.

Hardware

The hardware has three components:

- a set of processing elements
- a communication network
- a set of input/output processors and associated peripheral devices

The hardware architecture can be described as a hierarchical clustering of processing elements. Processing elements provide memory and processing resources, together with an interface to a communications system. The communications system facilitates the transfer of data between processing elements, and to and from peripheral devices. A single processing element is a cluster of level zero; the next two cluster levels contain 8, and 128, processing elements, respectively. The topology of interconnection of clusters forms a fractal network, analogous to a telephone network.

The communications bandwidth between clusters rolls off as one ascends the cluster hierarchy. For any given cluster, the ratio of the number of external communication links to the number of internal links is strictly less than unity. This makes possible a communications system which takes up a constant proportion of the total hardware resources, independent of the size of a hardware configuration. The decreasing communications bandwidth also implies that the cost, measured in latency, of communication between processing elements, increases as the latter are located further apart. Communication overheads are kept small by taking advantage of the locality of reference of programs written for virtual memory machines.

Control Firmware

The Myrias parallel computer is implemented as a virtual machine, called G, on which all user and user-visible software runs. G runs directly on the underlying hardware. The G machine is embodied in the control firmware. The term "firmware" is used in this context because it is intended that the control system will evolve slowly, thus presenting the user-level software with a firm foundation. A G machine, together with the hardware on which it runs, is called a G processor. The G machine has its own language, also called G.

G control firmware is transparent to the user of a Myrias parallel computer, and every effort is being undertaken to ensure that details of its implementation do not propagate out to the user programming model. The G control firmware performs three functions, and has three corresponding components.

The first component of the G control firmware is the G translator and run time system. The G translator accepts programs in the G language, and translates them into the native code of the underlying hardware. The run time system performs certain functions required by the translated code. A copy of the run time system resides in each processing element of a G processor.

The G Kernel, which is also totally distributed in a G processor, schedules parallel tasks on processing elements, merges the results of their execution, and provides virtual memory support to the G machine. During the execution of a program, the work load throughout processors is leveled by the Kernel. In particular, tasks may be redistributed by the Kernel if this is advantageous.

Virtual memory support involves management of all data motion, eg. arranging for the data required by parallel tasks to be available to the appropriate processing elements, when required. To this end, pages are cached at different levels of the hardware hierarchy. The G Kernel performs its functions in such a way as to optimize use of machine resources, and to minimize the overhead associated with data motion between processing elements.

The third component of the G control firmware is ROM based firmware that controls the communications subsystem of a G processor.

The system architecture is stable. It will be able to accommodate different implementations; for example with faster processors, larger and faster memories, and 64-bit addressing.

Operating Software

Operating software for the Myrias parallel computer runs on the virtual G machine. The operating software includes the user-visible operating system, a distributed form of AT&T Bell Laboratories' UNIX. In addition to UNIX, there are compilers, which accept user programs written in high-level parallel programming languages, and emit G language programs to be processed by the underlying G processor. Finally, there is a collection of tools required for linkage editing, loading, and debugging, together with mathematical subroutine libraries.

Programming Languages

At the very top of the multilayered machine come applications programs written in high-level programming languages. Myrias will support two high-level languages initially, Myrias Parallel Fortran (MPF; a superset of ANSI Standard Fortran 77), and Myrias Parallel C (MPC; a superset of the C language defined by UNIX System V). The languages are standard apart from three basic extensions discussed in § 3.

2.2 Present implementation of the architecture

The hardware implementation has been kept deliberately conservative to reduce development risk.

As mentioned above, a processing element (PE) combines memory, processing power, and communication facilities. In the present implementation, this memory is composed of 512 Kbytes DRAM, while the processing power is supplied by a Motorola MC68000 10 MHz processor.

A typical G configuration is intended to contain up to several thousand PEs. The scalable system will be available in "Krate" units of 1024 PEs each. Each Krate will consist of 8 card cages, where one card cage houses 16 multiple processing element boards. One such board includes 8 PEs, plus a supervisory processor and an interface to the backplane. Each card cage has provision for ports to 4 input/output processors, which control external devices. The scalability of the system implies that the amount of memory and number of processors is expandable, in principle indefinitely.

Primary memory capacity is 512 Mbytes per Krate. All memory in the Myrias System is distributed; there is no common memory. The system can be thought of as intelligent memory organized in the form of a hierarchical cache. The total memory-to-processor bandwidth is over 5 Gbytes/sec per Krate. This rate scales linearly with the number of Krates in a configuration. The hardware capability for throughput to external networks exceeds 200 Mbytes/sec per Krate.

Note: a 512 PE prototype G processor was built in 1986/1987. The production system is based on more capable hardware and will be available in 1989.

2.3 Performance considerations

System users will be assigned distinct domains in the system, where domains can consist of from one to thousands of processing elements. A single process running on a domain is inviolate.

The supervising G Kernel relieves the user from the very considerable burden of appropriate scheduling, load leveling, synchronization of parallel tasks, and tracking of communications among parallel processes throughout a code. The user will be able to program computer systems containing thousands of processing elements easily and efficiently.

The performance of a scalable Myrias system configuration will increase nearly linearly with its size, for sufficiently large tasks, exhibiting high parallelism. Performance will be good if there is good locality of reference among related parallel tasks. If a task must reference large numbers of widespread variables of little locality, however, system performance can be degraded as a result of excessive data motion.

3. New Memory Semantics and Fortran Extensions

The ability to model physical systems which involve large numbers of heterogeneous locales is fundamentally limited by the processing speed, the memory size, and the memory bandwidth of available computers.

The serial Fortran memory model amounts to a collection of storage units. Sequences of storage units are generally associated with variables and arrays. The normal flow of control in Fortran also implies a mechanism that keeps track of the current statement and of subroutine and function return points. Thus, the state of the Fortran machine at any point in time can be seen as a concatenation of the states of its constituent storage units, together with flow control information.

The new memory model discussed here allows several incarnations of the Fortran-programmed machine to coexist during execution. Each incarnation is called a *task*, and executes in its own separate memory space. These semantics provide the principal means to attain parallelism. In order to utilize parallelism within these memory semantics, it was necessary to develop languages that would match the parallel Myrias architecture [3]. Thus, 'Myrias Parallel Fortran' (MPF) and 'Myrias Parallel C' (MPC) have been developed.

The principal MPF language extensions comprise:

- the **par do** construct to generate concurrent execution of tasks within a **do** loop involving independent iterations
- recursive use of subroutine or function calls, providing the capability of parallel recursion
- dynamic array dimensioning, allowing multiple copies of array variables which are assigned at run-time instead of compile-time

The extensions are orthogonal, hence they may be combined freely. The ability to make recursive calls inside parallel loops is an extremely powerful and versatile tool, particularly in algorithms employing a *"divide-and-conquer"* approach [1,4].

With these extensions, many previously intractable computational problems can be solved easily. For example, it will be relatively straightforward and easy to employ fine grids and/or high precision arithmetic in regions of interest, while coarser grids and/or lower precision arithmetic can be used in other (asymptotic) regions. Moreover, the use of recursive refinement can follow the physics of a simulated process dynamically, thus very substantially reducing computational complexity (cf. ref [5,6].).

To allow for recursion within a parallel computing model, distributed memory with its own local workspace is a requirement which greatly simplifies possible complexities. A distributed memory eliminates the memory contention that one could face in a shared memory environment. Although a distributed memory system will always incur some form of overhead due to the splitting and merging of parallel tasks, for the most part, the costs can be relatively small compared to the gains from the parallelism invoked by the distribution of tasks. The only real constraint is that the work within each parallel task should be large enough to amortize the communication overhead.

In the Myrias system, in principle, each task gets a virtual copy of the current workspace, and manages its own portion of the program, totally oblivious to other concurrent tasks. Not until the tasks have all completed, and the merging of the parallel tasks takes place, does the necessity arise for much communication and its associated overheads. The advantage of a programming model that allows for recursion becomes apparent when recursive calls are made in parallel.

4. The PAR DO Extension

As an example, we show in detail how the **par do** (parallel do) extension employs the new memory semantics and thus achieves parallelism in the MPF programming model.

When a running task activates a **par do** loop, a number of new tasks are created, one for each iteration of the loop. Each iteration sees the machine state as it was at the beginning of the **par do** instead of as it was at the end of the previous loop iteration. Conceptually, the original task is the parent of many newly created child tasks. The child tasks may be completely heterogeneous and, conceptually, are done in parallel. Until all of its child tasks are completed or terminated, the parent task is suspended. The amount of actual parallelism is, of course, restricted by the number of processors in the domain. The scheduling is done by a firmware control system, not the user.

When a **par do** is invoked, the iteration count is established (just as for the serial **do**), and a new task is initiated for each iteration. The state of each new task is copied from the state of its parent task, with some minor changes to account for the loop control variable and control flow information.

A newly-created child task is executed until the terminal statement of the **par do** loop is encountered, whereupon the child task is completed. There is no communication among sibling tasks; each task is independent of the others.

Providing that there is no transfer of control out of range of a **par do**, once all of the child tasks have been completed, the state of the parent is updated using a *merging* rule. For each storage unit in the parent state, there are four possible cases:

(1) no update:

if no child task assigns to a variable, then the parent variable is unchanged.

(2) one task updates:

if exactly one child task assigns to a variable, the variable in the parent task is changed to the assigned value.

(3) several tasks update with the same value:

if more than one child task assign to a variable, but the values assigned are identical, then the variable in the parent task is changed to the assigned value.

(4) otherwise:

any other update pattern will cause the value of the parent task variable to be unpredictable. If several tasks assign different values to a variable, there is no natural way to chose which value should it have after merging.

The result of the merging process is that a new parent state is formed to incorporate results computed by its children. The parent task can then resume just as it would at the completion of a corresponding serial **do** loop.

An arbitration mechanism is used to identify the first such transfer of control with respect to a **par do**. The arbitration mechanism is implementation dependent.

MPF contains a semantic device to program the simultaneous execution of several distinct code blocks [7]. The code sequence

par begin
  code block 1
parallel
  code block 2
  . . .

```
  parallel
    code block n
  end par
is equivalent to
  par do 1 i = 1,n
    if i = 1 then
      code block 1
    elseif i = 2 then
      code block 2
      . . .
    elseif i = n then
      code block n
1   continue
```

Conceptually, all of the code blocks are executed in parallel.

Invariably, the system control software that implements the memory semantics avoids most of the copying implied by this conceptual description.

5. Parallel Processing in a Virtual Memory Environment

In many respects, the utilization of a distributed virtual memory creates new possibilities in parallel programming. A distributed cache, combined with a parallel programming model, creates new enhancements for a spectrum of important computational problems.

We describe here a few applications in which parallel processing is easily integrated into an environment such as the one discussed above. We also give some algorithmic examples which highlight the power of the new parallel programming model that allows for parallel loop execution and recursive subroutine and function calls.

5.1 Tridiagonal Linear Systems

We first describe an example from a typical linear algebra problem — the solution of a tridiagonal linear system. Tridiagonal and block-tridiagonal systems arise in numerous areas of physical and simulation modeling, usually from the use of finite difference methods with three-point difference schemes or, in the case of block tridiagonals, with higher order difference schemes, applied to solve ordinary and partial differential equations (cf. refs. [8, 9]).

Although the solution to a tridiagonal linear system can be easily vectorized by placing the diagonals in contiguous arrays, we show the power of a **par do** by describing a recursive parallel method. A powerful parallel solution to the problem is to solve the system by the Odd-Even Reduction or Odd-Even Elimination algorithms [1, 10, 11].

The method analyzed here, a solution by odd-even elimination, allows, as shown below, for the system to be solved in $O(\log_2 n)$ steps. The basic idea of parallel odd-even elimination accommodates a twofold parallelism. First of all, the reduction of odd-indexed elements in even-indexed rows can be done concurrently with the reduction of even-indexed elements in odd-indexed rows. Secondly, parallelism exists in the elimination of elements in one row; the elements to the left of the main diagonal can be eliminated independent of the elements to the right of the main diagonal. Therefore, while the left off-diagonal element of row i is being eliminated by the multiplication of the main diagonal element of row i-1, the right off-diagonal element of row i can be eliminated by the element in the main diagonal of row i+1. And since each row i only needs the current values of rows i-1 and i+1, all rows can be done in parallel. The only complication arises from the need for the multipliers generated by the elimination of the preceding and the following rows. This problem, however, can be solved by simply storing all multipliers in temporary arrays and adding them to the appropriate diagonal elements after merging the parallel tasks. Therefore, after one parallel step, the linear system will be adjusted so that the off-diagonal elements are "pushed out" to the left and right. Graphically, this effect can be depicted in the following way:

We assume a system $A \cdot \mathbf{x} = \mathbf{y}$ such that A is a 7 x 7 tridiagonal matrix.

$$A = \begin{pmatrix} x & x & & & & & \\ x & x & x & & & & \\ & x & x & x & & & \\ & & x & x & x & & \\ & & & x & x & x & \\ & & & & x & x & x \\ & & & & & x & x \end{pmatrix}_{7 \times 7}$$

where x's represent non-zero elements.

After the first step of the elimination process, the distribution of non-zero elements would be:

$$A = \begin{pmatrix} x & & x & & & & \\ & x & & x & & & \\ x & & x & & x & & \\ & x & & x & & x & \\ & & x & & x & & x \\ & & & x & & x & \\ & & & & x & & x \end{pmatrix}_{7 \times 7}$$

After the second step, the elements would be further pushed out:

$$A = \begin{pmatrix} x & & & & x & & \\ & x & & & & x & \\ & & x & & & & x \\ & & & x & & & \\ x & & & & x & & \\ & x & & & & x & \\ & & x & & & & x \end{pmatrix}_{7 \times 7}$$

Thus, if we recurse and repeat at each step, we can eventually eliminate all off-diagonal elements, until only the main diagonal remains. We also note that, through each level of recursion in the elimination process, the position of the remaining off-diagonal elements is moved out by twice the previous spacing. Hence, the complete system can be computed, in parallel, in only $O(\log_2 n)$ steps.

For convenience of efficient storage, we assume a tridiagonal matrix which has elements stored in such a way that the main diagonal is in column 2, the left and right off-diagonals are in columns 1 and 3 respectively, and the right-hand-side elements are in column 4. For such a case, we present here a subroutine which does parallel odd-even elimination for just such a tridiagonal matrix. The routine will reduce a tridiagonal matrix, 'mat', to a matrix with only a main diagonal. In order to illustrate both the algorithm and the programming language, we give the full code, written in Myrias Parallel Fortran, as follows:

```
      k = 1
      n = no_of_rows

      subroutine TRI_ELIM (mat, n, k)
      integer n, k
      real mat(n,4), ltmp(n), rtmp(n), lbtmp(n), rbtmp(n), um, lm
      if (k .lt. n) then
        par begin

*   Left off-diagonals are computed with a lower multiplier (lm)

          par do 10 i = k, n-k
            lm = - mat(i+k,1) / mat(i,2)
            mat(i+k,1) = lm * mat(i,1)
            ltmp(i+k) = lm * mat(i,3)
            lbtmp(i+k) = lm * mat(i,4)
10        continue
        parallel

*   In parallel, right off-diagonals are computed with an upper multiplier (um)

          par do 20 j = k+1, n
            um = - mat(j-k,3) / mat(j,2)
            mat(j-k,3) = um * mat(j,3)
            rtmp(j-k) = um * mat(j,1)
            rbtmp(j-k) = um * mat(j,4)
20        continue
        end par

*  Here the results are merged in parallel by adding the stored elements
*  to the appropriate diagonal and right-hand-side elements

        par begin
          par do 30 i = 1, k
            mat(i,2) = mat(i,2) + rtmp(i)
            mat(i,4) = mat(i,4) + rbtmp(i)
30        continue
        parallel
          par do 40 i = k+1, n-k
            mat(i,2) = mat(i,2) + ltmp(i) + rtmp(i)
            mat(i,4) = mat(i,4) + lbtmp(i) + rbtmp(i)
40        continue
        parallel
          par do 50 i = (n-k) + 1, n
            mat(i,2) = mat(i,2) + ltmp(i)
            mat(i,4) = mat(i,4) + lbtmp(i)
```

```
50      continue
      end par
      k = k * 2
      call TRI_ELIM (mat, n, k)
    endif
    end
```

The potential gains of this algorithm can be substantial if the system is extrapolated to a block-tridiagonal linear system. To extend the above algorithm to the block-tridiagonal case is trivial, requiring the simple extension of block matrices in lieu of individual elements throughout the diagonals. Since many numerical modeling problems involve solutions of such systems, the importance of parallelism is easily recognized.

We note that block-tridiagonal systems resist effective vectorization since vector lengths achieved are usually as short as the block size.

The computational advantages of these algorithms result from the twofold embedded parallelism implicit in them. Given a large problem, and a sufficiently large computer domain size (i.e. number of processors) to work with, computations of this type are very amenable to a parallel computer with the programming model described above. Theoretical analyses indicate that with a scalable system, for large matrices, the speed-up (over serial execution of the parallel algorithm) will be almost proportional to the number of processors employed.

5.2 Adaptive Grids in Aerospace Applications

Some of the most computationally-intense problems, for which supercomputers are a necessity, are in aerodynamics, where partial differential equations are solved under various approximations in order to simulate the air-flow around an aircraft in flight [9]. A common method is to use a two- or three-dimensional computational grid, and compute values of flow at each of the grid-points. In order to do realistic simulations of sufficient accuracy, a large number of well defined grid-points are required, leading to immense computations [12,13].

Most interesting simulation problems are highly inhomogeneous or irregular, requiring adaptive, high order algorithms for their solution [5,6]. For example in transonic flow, the supersonic and subsonic regions of flow are described by different types of differential equations. In addition, it is desirable to expand computations in regions of high density gradients, where shock waves are formed, since it is important to track these regions more accurately than the smooth, asymptotic regions. This can be effectively achieved by recursive refinements of the computational grid in regions where physical gradients are large, and keeping it coarse where gradients are small. Moreover, such grid adaptation can proceed dynamically, when required by evolution of the flow. Parallel adaptive grid techniques can substantially decrease execution times with no loss of accuracy in the interesting regions, and at the same time, allow for a substantial increase in the complexity of flows.

A classical, Jacobi approach to the problem of adaptive grids, in Myrias Parallel Fortran, might have the following heuristic form:

```
do 4 n = 1, nsteps        step through time

   par do 3 i = 1, nx        step through a
      par do 2 j = 1, ny        three dimensional
         par do 1 k = 1, nz        grid
```

```
          if density gradient at (i,j,k) is small then
             if point (i,j,k) is supersonic then

                Compute next flow value at point (i,j,k) using
                elliptic difference operator

             elseif  point (i,j,k) is subsonic then

                Compute next flow value at point (i,j,k) using
                hyperbolic difference operator

             endif

          else density gradient at (i,j,k) is large

               Subdivide the grid recursively until the mesh is
               fine enough to give the desired accuracy, then
               compute the next flow values using a finer time step

          endif
1          continue
2        continue
3      continue
4  continue
```

Through the use of dynamic array allocation, the Myrias system is able to reconfigure itself in places of high interest, allowing more memory space to be utilized in the grid areas which require more computations. Combined with parallel recursion, such utilization of memory and parallelism could greatly shorten overall computational time.

6. Summary

We have described both a powerful new memory model for parallel computation, and the Myrias architectural model on which it is being implemented. In its entirety, this model is expected to lead to significant improvements in programming complexity, computational times, and, in general, in numerical physical modeling. Combined with a virtual memory model that utilizes local, distributed memory, the Myrias parallel computing system allows for natural algorithmic concepts to be integrated into a parallel computing environment.

We have also described some of the details of the memory organization, the virtual *G* machine, and the hardware substrate on which it runs. Moreover, we have discussed a few of the many important applications for which such a computer system will be able to shorten computational times.

7. References

1. D. Heller, A Survey of Parallel Algorithms in Numerical Linear Algebra, *SIAM Review 20,* (1978), 740-777.
2. *Myrias 4000 System Description (unpublished),* Myrias Research Corp., Edmonton, Alberta, Feb. 1986.

3. J. R. Savage, Parallel Processing as a Language Design Problem, *Proc. 12th Annual Int. Symp. on Computer Architecture*, Boston, MA, 1985, 221-224.

4. C. Y. Tang, Optimal Speeding-up of Parallel Algorithms Based Upon the Divide-and-Conquer Strategy, *Information Sciences 32*, (1984), 173-186.

5. J. Ortega and R. Voigt, Solution of Partial Differential Equations on Vector and Parallel Computers, *SIAM Review 27*, (1985), 149-240.

6. D. Gannon and J. Van Rosendale, Parallel Architectures for Iterative Methods on Adaptive, Block Structured Grids, in *Elliptic Problem Solvers II*, 1984, 93-104.

7. P. O. Frederickson, R. E. Jones and B. T. Smith, Synchronization and control of parallel algorithms, *Parallel Computing 2*, (1985), 255-264, North-Holland.

8. J. M. Varah, On the Solution of Block-Tridiagonal Systems Arising from Certain Finite-Difference Equations, *Mathematics of Computation 26*, 120 (1972), 859-868.

9. W. Gentzsch, Numerical algorithms in computational fluid dynamics on vector computers, *Parallel Computing 1*, (1984), 19-33, North-Holland.

10. D. Heller, Some Aspects of the Cyclic Reduction Algorithm for Block Tridiagonal Linear Systems, *SIAM J. Numer. Anal. 13*, 4 (1976), 484-496.

11. R. N. Kapur and J. C. Browne, Techniques for Solving Block Tridiagonal Systems on Reconfigurable Array Computers, *SIAM Journal of Scientific Statistical Computing 5*, (1984), 701-719.

12. F. R. Bailey, *Computational Limits on Scientific Applications*, IEEE Proceedings of Compcon Spring, 1978.

13. V. L. Peterson, W. F. Ballhaus and F. R. Bailey, Numerical Aerodynamical Simulation (NAS), *NASA Technical Memorandum 84386*, Moffet Field, CA, July 1983.

INDEX